D1521613

QUANTUM ALGORITHMS VIA LINEAR ALGEBRA

QUANTUM ALGORITHMS VIA LINEAR ALGEBRA

A Primer

Richard J. Lipton
Kenneth W. Regan

The MIT Press
Cambridge, Massachusetts
London, England

MIT Press books may be purchased at special quantity discounts for business or sales promotional use. For information, please email special_sales@mitpress.mit.edu.

This book was set in Times Roman and Mathtime Pro 2 by the authors, and was printed and bound in the United States of America.

Library of Congress Cataloging-in-Publication Data

Lipton, Richard J., 1946–
Quantum algorithms via linear algebra: a primer / Richard J. Lipton and Kenneth W. Regan.
p. cm.
Includes bibliographical references and index.
ISBN 978-0-262-02839-4 (hardcover : alk. paper)
1. Quantum computers. 2. Computer algorithms. 3. Algebra, Linear. I. Regan, Kenneth W., 1959–
II. Title

QA76.889.L57 2014
005.1–dc23

2014016946

10 9 8 7 6 5 4 3

We dedicate this book to all those who helped create and nourish the beautiful area of quantum algorithms, and to our families who helped create and nourish us.

RJL and KWR

Contents

Preface

This book is an introduction to quantum algorithms unlike any other. It is short, yet it is comprehensive and covers the most important and famous quantum algorithms; it assumes minimal background, yet is mathematically rigorous; it explains quantum algorithms, yet steers clear of the notorious philosophical problems and issues of quantum mechanics.

We assume no background in quantum theory, quantum mechanics, or quantum anything. None. Quantum computation can be described in terms of elementary linear algebra, so some familiarity with vectors, matrices, and their basic properties is required. However, we will review all that we need from linear algebra, which is surprisingly little. If you need a refresher, then our material should be enough; if you are new to linear algebra, then we suggest some places where you can find the required material. It is really not much, so do not worry.

We do assume that you are comfortable with mathematical proofs; that is, we assume "mathematical maturity" in a way that is hard to define. Our proofs are short and straightforward, except for advanced topics in section 13.5 and chapters 15 and 16. This may be another surprise: for all the excitement about quantum algorithms, it is interesting that the mathematical tools and methods used are elementary. The proofs are neat, clever, and interesting, but you should have little trouble following the arguments. If you do, it is our fault—we hope that our explanations are always clear. Our idea of a standard course runs through section 13.4, possibly including chapter 14.

We strive for mathematical precision. There is always a fine line between being complete and clear and being pedantic—hopefully we stay on the right side of this. We started with the principle of supplying all the details—all of them—on all we present. We have compromised in three places, all having to do with approximations. The first is our using the quantum Fourier transform "as-is" rather than approximating it, and the others are in chapters 15 and 16.

For better focus on the *algorithms*, we chose to de-emphasize quantum *circuits*. In fact, we tried to avoid quantum circuits and particularities of quantum gates altogether. However, they are excellent to illuminate linear algebra, so we have provided a rich set of exercises in chapters 3 through 7, plus two popular applications in section 8.3. These can in fact be used to support coverage of quantum circuits in a wider-scale course. The same goes for complexity classes. We prefer to speak operationally in terms of *feasible* computation, and we try to avoid being wedded to the "asymptotically polynomially bounded" definition of it. We avoid *naming* any complexity class until chapter 16. Nevertheless, that chapter has ample complexity content anchored in computational problems rather than machine models and is self-contained enough to support a course that covers complexity theory. At the same stroke, it gives algebraic tools for analyzing quantum circuits. We featured tricks we regard as *algorithmic* in the main text and delegated some tricks of implementation to exercises.

What makes an algorithm a *quantum* algorithm? The answer should have nothing to do with how the algorithm is implemented in a physical quantum system. We regard this as really a question about how programming notation—mathematical notation—represents the feasibility of calculations in nature. Quantum algorithms use algebraic units called *qubits* that are richer than bits, by which they are allowed to count as feasible some operations that when written out in simple linear algebra use exponentially long notation. The rules for these allowed operations are specified in standard models of quantum computation, which are all equivalent to the one presented in this book. It might seem ludicrous to believe that nature in any sense *uses* exponentially long notation, but some facet of this appears at hand because quantum algorithms can quickly solve problems that many researchers believe require exponential work by *any* "classical" algorithm. In this book, classical means an algorithm written in the notation for feasible operations used by every computer today.

This leads to a word about our presentation. Almost all summaries, notes, and books on quantum algorithms use a special notation for vectors and matrices. This is the famous Dirac notation that was invented by Paul Dirac—who else. It has many advantages and is the de-facto standard in the study of quantum algorithms. It is a great notation for experts and instrumental to becoming an expert, but we suspect it is a barrier for those starting out who are not experts. Thus, we avoid using it, except for a few places toward the end to give a second view of some complicated states. Our thesis is that we can explain quantum algorithms without a single use of this notation. Essentially this book is a testament to that belief: if you find this book more accessible than others, then we believe it owes to this decision. Our notation follows certain ISO recommendations, including boldface italics for vectors and heavy slant for matrices and operators.

We hope you will enjoy this little book. It can be used to gain an understanding of quantum algorithms by self-study, as a course or seminar text, or even as additional material in a general course on algorithms.

Georgia Institute of Technology, *Richard J. Lipton*
University at Buffalo (SUNY), *Kenneth W. Regan*

Acknowledgements

We thank Aram Harrow, Gil Kalai, and John Preskill for contributions to a debate since 2012 on themes reflected in the opening and last chapters, Andrew Childs and Stacey Jeffery for suggestions on chapter 15, and several members of the IBM Thomas J. Watson Research Center for suggestions and illuminating discussions that stimulated extra coverage of quantum gates and circuit simulations in the exercises. We thank Marie Lufkin Lee and Marc Lowenthal and others at MIT Press for patiently shepherding this project to completion, and the anonymous reviewers of the manuscript in previous stages. We also thank colleagues and students for some helpful input and thank others—most notably Cem Say and his group at Boğaziçi University in Ankara—for noting errata in the first printing. Quantum circuit diagrams were typeset using version 2 of the `Qcircuit.tex` package by Steve Flammia and Bryan Eastin.

1 Introduction

One of the great scientific and engineering questions of our time is:

Are quantum computers possible?

We can build computers out of mechanical gears and levers, out of electric relays, out of vacuum tubes, out of discrete transistors, and finally today out of integrated circuits that contain thousands of millions of individual transistors. In the future, it may be possible to build computers out of other types of devices—who knows.

All of these computers, from mechanical to integrated-circuit-based ones, are called *classical*. They are all classical in that they implement the same type of computer, albeit as the technology gets more sophisticated the computers become faster, smaller, and more reliable. But they all behave in the same way, and they all operate in a non-quantum regime.

What distinguishes these devices is that information is manipulated as *bits*, which already have determinate values of 0 or 1. Ironically, the key components of today's computers are quantum devices. Both the transistor and its potential replacement, the Josephson junction, won a Nobel Prize for the quantum theory of their operation. So why is their regime non-quantum? The reason is that the regime reckons information as bits.

By contrast, quantum computation operates on *qubits*, which are based on complex-number values, not just 0 and 1. They can be read only by *measuring*, and the readout is in classical bits. To skirt the commonly bandied notion of observers interfering with quantum systems and postpone the discussion of measurement as an operation, we offer the metaphor that a bit is what you get by "cooking" a qubit. From this standpoint, doing a classical computation on bits is like cooking the ingredients of a pie individually before baking them together in the pie. The quantum argument is that it's more expedient to let the filling bubble in its natural state while cooking everything at once. The engineering problem is whether the filling can stay *coherent* long enough for this to work.

The central question is whether it is possible to build computers that are inherently quantum. Such computers would exploit the power and wonder of nature to create systems that can effectively be in multiple states at once. They open a world with apparent actions at a distance that the great Albert Einstein never believed but that actually happen—a world with other strange and counter-intuitive effects. To be sure, this is the world we live in, so the question becomes how much of this world our computers can enact.

This question is yet to be resolved. Many believe that such machines will be built one day. Some others have fundamental doubts and believe there are physical limits that make quantum computers impossible. It is currently unclear who is right, but whatever happens will be interesting: a world with quantum computers would allow us to solve hard problems, while a barrier to them might shed light on deep questions of physics and information.

Happily this question does not pose a barrier to us. We plan to study **quantum algorithms**, which are interesting whether quantum computers are built soon, in the next ten years, in the next fifty years, or never. The area of quantum algorithms contains some beautiful ideas that everyone interested in computation should know.

The rationale for this book is to supply a gentle introduction to quantum algorithms. We will say nothing more about quantum computers—about whether they will be built or how they may work—until the end. We will only discuss algorithms.

Our goal is to explain quantum algorithms in a way that is accessible to almost anyone. Curiously, while quantum algorithms are quite different from classical ones, the mathematical tools needed to understand them are quite modest. The mathematics that is required to understand them is linear algebra: vectors, matrices, and their basic properties. That is all. So these are really linear-algebraic algorithms.

1.1 The Model

The universe is complex and filled with strange and wonderful things. From lifeforms like viruses, bacteria, and people; to inanimate objects like computers, airplanes, and bridges that span huge distances; from planets, to whole galaxies. There is mystery and wonder in them all.

The goal of science in general, and physics specifically, is to explore and understand the universe by discovering the simplest laws possible that explain the multitude of phenomena. The method used by physics is the discovery of models that predict the behavior of all from the smallest to the largest objects. In ancient times, the models were crude: the earliest models "explained" all by reducing everything to earth, water, wind, and fire. Today, the models are much more refined—they replace earth and water by hundreds of particles and wind and fire by the four fundamental forces. Mainly, the models are better at predicting, that is, in channeling *reproducible knowledge*. Yet the full theory, the long-desired theory of everything, still eludes us.

Happily, in this introduction to quantum algorithms, we need only a simple model of how part of the universe works. We can avoid relativity, special and general; we can avoid the complexities of the Standard Model of particle physics, with its hundreds of particles; we can even avoid gravity and electromagnetism. We cannot quite go back to earth, water, wind, and fire, but we can avoid having to know and understand much of modern physics. This avowal of independence from physical qualities does not prevent us from imagining nature's workings. Instead, it speaks to our belief that algorithmic considerations in information processing run deeper.

1.2 The Space and the States

So what do we need to know? We need to understand that the state of our quantum systems will always be described by a single unit vector $\mathbf{a}$ that lies in some fixed vector space of dimension $N = 2^n$, for some n. That is, the state is always a vector

$$\mathbf{a} = \begin{bmatrix} a_0 \\ \vdots \\ a_{N-1} \end{bmatrix},$$

where each entry a_k is a real or complex number depending on whether the space is real or complex. Each entry is called an **amplitude**. We will not need to involve the full quantum theory of *mixed states*, which are formally the same as classical probability distributions over states like $\mathbf{a}$, which are called *pure states*. That is, we consider only pure states in this text.

We must distinguish between general states and **basis states**, which form a linear-algebra basis composed of configurations that we may observe. In the **standard basis**, the basis states are denoted by the vectors $\mathbf{e}_k$ whose entries are 0 except for a 1 in place k. We identify $\mathbf{e}_k$ with the index k itself in $[0, N-1]$ and then further identify k with the k-th string x in a fixed total ordering of $\{0,1\}^n$. That the basis states correspond to all the length-n binary strings is why we have $N = 2^n$. The interplays among basis vectors, numerical indices, and binary strings encoding objects are drawn more formally in chapter 2.

Any vector that is not a basis state is a **superposition**. Two or more basis states have nonzero amplitude in any superposition, and only one of them can be *observed* individually in any *measurement*. The amplitude a_k is not directly meaningful for what we may expect to observe but rather its squared absolute value, $|a_k|^2$. This gives the *probability* of *measuring* the system to be in the basis state $\boldsymbol{e}_k$. The import of $\boldsymbol{a}$ being a unit vector is that these probabilities sum to 1, namely, $\sum_{k=0}^{N-1} |a_k|^2 = 1$.

For $N = 2$, the idea that the length of a diagonal line defined by the origin and a point in the plane involves the sum of two squares is older than Pythagoras. The length is 1 precisely when the point lies on the unit circle. We may regard the basis state $\boldsymbol{e}_0$ as lying on the x-axis, while $\boldsymbol{e}_1$ lies on the y-axis. Then measurement projects the state $\boldsymbol{a}$ either onto the "x leg" of the triangle it makes with the x-axis or the "y leg" of the triangle along the y-axis.

It may still seem odd that the probabilities are proportional not to the lengths of the legs but to their squares. But we know that if the angle is θ from the near part of the x-axis (so $0 \leq \theta \leq \pi/2$), then the lengths are $\cos(\theta)$ and $\sin(\theta)$, respectively, and it is $\cos^2(\theta) + sin^2(\theta)$, not $\cos(\theta) + \sin(\theta)$, that sums to 1. If we wanted to use points whose legs sum directly to 1, we'd have to use the diamond that is inscribed inside the circle. In N-dimensional space, we'd have to use the N-dimensional *simplex* rather than the sphere. Well the simplex is spiky and was not really studied until the 20th century, whereas the sphere is smooth and nice and was appreciated by the ancients. Evidently nature agrees with ancient aesthetics. We may not know *why* the world works this way, but we can certainly say, *why not?*

Once we agree, all we really need to know about the space is that it supports the picture of Pythagoras, that is, ordinary Euclidean space. Both the real vector spaces $\mathbb{R}^N$ and the complex ones $\mathbb{C}^N$ do so. That is, they agree on how many components their vectors have and how distances are measured by taking squares of values from the vector components. They differ only on what kind of numbers these component values v can be, but the idea of the norm or absolute value $|v|$ quickly reconciles this difference. This aspect was first formalized by Euclid's great rigorizer of the late 19th and early 20th centuries, David Hilbert; in his honor, the common concept is called a **Hilbert Space**. Hilbert's concept retains its vigor even if "N" can be infinite, or if the "vectors" and "components" are strange objects, but in this book, we need not worry: N will always be finite, and the space $\mathbb{H}^N$ will be $\mathbb{R}^N$ or $\mathbb{C}^N$. Allowing the latter is the reason that we say "Hilbert space" not "ordinary Euclidean space."

1.3 The Operations

In fact, our model embraces the circle and the sphere even more tightly. It throws out—it disallows—all other points. Every state $\boldsymbol{a}$ must be a point on the unit sphere. If you have heard the term "projective space," then that is what we are restricting to. But happily we need not worry about restricting the *points* so long as we restrict the *operations* in our model. The operations must map any point on the unit sphere to some (other) point on the unit sphere.

We will also restrict the operations to be *linear*. Apart from *measurements*, they must be onto the whole sphere—which makes them map all of $\mathbb{H}^N$ onto all of $\mathbb{H}^N$. By the theory of vector subspaces, this means the operations must also be invertible. Operations with all these properties are called **unitary**.

We will represent the operations by matrices, and we give several equivalent stipulations for **unitary matrices** in chapter 3, followed by examples in chapter 5 and tricks for working with them in chapter 6. But we can already understand that compositions of unitary operations are unitary, and their representations and actions can be figured by the familiar idea of multiplying matrices. Thus, our model's *programs* will simply be compositions of unitary matrices. The one catch is that the matrices themselves will be huge, out of proportion to the actual simplicity of the operation as we believe nature meters it. Hence, we will devote time in chapters 4 and 5 to ensuring these operations are **feasible** according to standards already well accepted in classical computation.

Thus, we can comply with the requirements for any computational model that we must understand what state the computation starts in, how it moves from one state to another, and how we get information out of the computation.

Start: We will ultimately be able to assume that the start state is always the elementary vector

$$\boldsymbol{e}_0 = \begin{bmatrix} 1 \\ 0 \\ \vdots \\ 0 \end{bmatrix}$$

of length N. Because the first binary string in our ordering of $\{0, 1\}^n$ will be 0^n, our usual start state will denote the binary string of n zeros.

Move: If the system is in some state $\boldsymbol{a}$, then we can move it by applying a unitary transformation $\boldsymbol{U}$. Thus, $\boldsymbol{a}$ will move to $\boldsymbol{b}$ where $\boldsymbol{b} = \boldsymbol{U}\boldsymbol{a}$. Not all unitary transformations are allowed, but we will get to that later. Note that if $\boldsymbol{a}$ is a unit vector, then so is $\boldsymbol{b}$.

End: We get information out of the quantum computation by making a **measurement.** If the final state is $\boldsymbol{c}$, then k is seen with probability $|c_k|^2$. Note that the output is just the index k, *not* the probability of the index. Often we will have a distinguished set S of indices that stand for *accept*, with outcomes in $[0, N-1] \setminus S$ standing for *reject*.

That is it.

1.4 Where Is the Input?

In any model of computation, we expect that there is some way to input information into the model's devices. That seems, at first glance, to be missing from our model. The start state is fixed, and the output method is fixed, so where do we put the input? The answer is that the input can be encoded by the choice of the unitary transformation $\boldsymbol{U}$, in particular by the first several unitary operations in $\boldsymbol{U}$, so that for different inputs we will apply different transformations.

This can easily be done in the case of classical computations. We can dispense with explicit input provided we are allowed to change the program each time we want to solve a different problem. Consider a program of this form:

```
M = 21;
x = Factor(M);

procedure Factor(z) {  ...  }
```

Clearly, if we can access the value of the variable x, then we can determine the factors of 21. If we wanted to factor a more interesting number such as 35 or 1,001 or 11,234,143, then we can simply execute the same program with M set to that number.

This integration of inputs with programs is more characteristic of quantum than classical computation. Think of the transformation $\boldsymbol{U}$ as the program, so varying $\boldsymbol{U}$ is exactly the same as varying the above program in the classical case. We will show general ways of handling inputs in chapter 6, while for several famous algorithms, in chapters 8–10, the input is expressly given as a transformation that is dropped into a larger program. Not to worry—our chapter 7 in the middle also provides support for the classical notion of feeding a binary string x directly as the input, while also describing the ingredients of the "quantum power" that distinguish these programs from classical ones.

1.5 What Exactly Is the Output?

The problematic words in this section title are the two short ones. With apologies to President Bill Clinton, the question is not so much what the meaning of "is" is, as what the meaning of "the" is. In a classical deterministic model of computation, every computation has one, definite, output. Even if the model is randomized, every random input still determines a single computation path, whose output is definite before its last step is executed.

A quantum computation, however, presents the user at the end with a slot machine. Measuring is pulling the lever to see what output the wheels give you. Unlike in some old casinos where slot machines were rigged, you control how the machine is built, and hopefully you've designed it to make the probabilities work in your favor. As with the input, however, the focus shifts from a given binary string to the machinery itself, and to the action of *sampling a distribution* that pulling the measurement lever gives you.

In chapter 6, we will also finesse the issue of having measurements in the middle of computations that continue from states projected down to a lower-dimensional space. This could be analogized to having a slot machine still spinning after one wheel has fixed its value. We show why measurements may generally be postponed until the end, but the algorithms in this textbook already behave that way. This helps focus the idea that the ultimate goal of quantum computation is not a single output but rather a *sampling device*. Chapter 6 also lays groundwork for how those devices can be *re-used* to improve one's chance of success. Chapter 7 lays out the format for how we present and analyze quantum algorithms and gives further attention to entanglement, interference, and measurement.

All of this remains only philosophy, however, unless we can show how the results of the sampling help solve concrete problems efficiently in distinguished ways. These solutions are the ultimate outputs, as exemplified in chapters 8–10. In chapters 11 and 12, the outputs are factorizations of numbers, via the algorithm famously discovered by Peter Shor. In chapters 13–15, they are objects that we need to search for in a big search space. Chapter 16 branches out to topics in quantum complexity theory, defining the class BQP formally and proving upper and lower bounds on it in terms of classical complexity. Chapter 17 summarizes the algorithms and discusses some further topics and readings. Saying this completes the application of the model and the overview of this book.

1.6 Summary and Notes

In old Platonic fashion, we have tried to idealize quantum computation by pure thought apart from physical properties of the world around us. Of course one can make errors that way, as Aristotle showed by failing to climb a tower and drop equal-sized balls of unequal weights. It is vain to think that Pythagoras or Euclid or Archimedes could have come up with such a computational model but maybe not so vain to estimate it of Hilbert. Linear algebra and geometry were both deepened by Hilbert. Meanwhile, the quantum theory emerged and ideas of *computation* were extensively discussed long before Alan Turing gave his definitive classical answer (Turing, 1936).

The main stumbling block may have been probability. Quantum physicists were forced to embrace probability from the get-go, in a time when Newtonian determinism was dominant and Einstein said "God"—that is, nature—"does not play dice." Some of our algorithms will be deterministic—that is, we will encounter cases where the final points on the unit sphere coincide with standard basis vectors, whereupon all the probabilities are 0 or 1. However, coping with probabilistic output appears necessary to realize the full power of quantum computation. Another block, even for the physicists happy with dice, may have been the long time it took to realize the essence of computation. Turing's paper reached full flower only with the emergence of computing machines during and after World War II.

Even then, it took Richard Feynman, arguably the main visionary in the area after the passing of John von Neumann in 1956, until his last decade in the 1980s to set his vision down (Feynman, 1982, 1985). That is when it came to the attention of David Deutsch (1985) and some others. The theory still had several false starts—for instance, the second of us overlapped with Deutsch in 1984–1986 at Oxford's Mathematical Institute and saw its fellows turn aside Deutsch's initial claims to be able to compute classically uncomputable functions.

It can take a long time for a great theory to mature, but a great theorem such as Peter Shor's on quantum factoring can accelerate it a lot. We hope this book helps make the ascent to understanding it easier. We chose $M = 21$ in section 1.4 because it is currently the highest integer on which practical runs of Shor's algorithm have been claimed, but even these are not definitively established (Smolin et al., 2013).

2 Numbers and Strings

Before we can start to present quantum algorithms, we need to discuss the interchangeable roles of natural numbers and Boolean strings. The set $\mathbb{N}$ of *natural numbers* consists of

$$0, 1, 2, 3, \ldots$$

as usual.

A *Boolean string* is a string that consists solely of *bits*: a bit is either 0 or 1. In computer science, such strings play a critical role, as you probably already know, because our computational devices are all based on being either "on" or "off," "charged" or "uncharged," "magnetized" or "unmagnetized," and so on.

The operations we use on natural numbers are the usual ones. For example, $x + y$ is the sum of x and y, and $x \cdot y$ is their product. There is nothing new or surprising here. The operations we use on Boolean strings are also quite simple: The length of a string is the number of bits in the string, and if x and y are Boolean strings, then xy is their concatenation. Thus, if $x = 0101$ and $y = 111$, then we have $xy = 0101111$. This is a kind of "product" operation on strings, but we find it convenient not to use an explicit operator symbol. If you see xy and both x and y are strings, then xy is the result of concatenating them together.

What we need to do is switch from numbers to Boolean strings and back. Sometimes it is best to use the number representation and other times the string representation. This kind of dual nature is basic in computer science and will be used often in describing the quantum algorithms. There are, however, some hitches that must be regarded to make it work properly. Let's look and see why.

If m is a natural number, then it can uniquely be written as a binary number: Let

$$m = 2^{n-1}x_{n-1} + \cdots + 2x_1 + x_0,$$

where each x_i is a bit, and we insist that x_{n-1} is nonzero. Then we can use m to denote the Boolean string $x_{n-1} \ldots x_1 x_0$. For instance, 7 maps to the string 111. In the reverse direction, we can use the string

$$x_{n-1}, \ldots, x_0$$

to denote the natural number

$$2^{n-1}x_{n-1} + \cdots + 2x_1 + x_0.$$

For example, the string 10010 is the number $16 + 2 = 18$.

Often we will be concerned with numbers in a fixed range $0 \ldots N-1$ where $N = 2^n$. This range is denoted by $[N]$. Then it will be convenient to omit the leading 1 and use n bits for each number, so that zero is n-many 0s, one is $0^{n-1}1$, and so on up to $N-1=1^n$, where 1^n means n-many 1's. We call this the **canonical numbering of** $\{0,1\}^n$. For example, with $n=3$:

$$\begin{array}{rclrcl} 000 & = & 0 & 100 & = & 4 \\ 001 & = & 1 & 101 & = & 5 \\ 010 & = & 2 & 110 & = & 6 \\ 011 & = & 3 & 111 & = & 7. \end{array}$$

The small but important issue is that, for the representation from numbers to strings to be unambiguous, we must know how long the strings are. Otherwise, what does the number 0 represent? Does it represent 0 or 00 or 000 and so on? This is why we said earlier that the mapping between numbers and strings is not exact. To make it precise, we need to know the length of the strings. A more technical way of saying this is that once we specify the mapping as being between the natural numbers $0,1,\ldots,2^n-1$ and the strings of length n (that is, $\{0,1\}^n$), it is one-to-one. Note that 0 as a number now corresponds to the unique string

$$\underbrace{0\ldots0.}_{\text{total of } n \text{ zeros}}$$

There is one more operation that we use on Boolean strings. If x and y are Boolean strings of length m, then $x \bullet y$ is their **Boolean inner product**, which is defined to be

$$x_1y_1 \oplus \cdots \oplus x_my_m.$$

Here $\oplus$ means exclusive-or, which is the same as addition modulo 2. Hence, sometimes we may talk about Boolean strings as being members of an m-dimensional space with addition modulo 2. We must warn that the name inner product is also used when we talk about Hilbert spaces in chapter 3. Many sources use $x \cdot y$ to mean concatenation of strings, but we reserve the lighter dot for numerical multiplication. When x and y are single bits, $x \cdot y$ is the same as $x \bullet y$, but using the lighter dot still helps remind us that they are single bits. Sometimes this type of overloading occurs in mathematics—we try to make clear which is used when.

A further neat property of Boolean strings is that they can represent subsets of a set. If the set is $\{1,2,3\}$, in that order, then 000 corresponds to the empty set, 011 to $\{2,3\}$, 100 to $\{1\}$, 111 to the whole set $\{1,2,3\}$, and so on.

2.1 Asymptotic Notation

Suppose we run an algorithm that on problems of size n works in n "passes," where the i-th pass takes i "steps." How long does the whole algorithm take? If we want to be exact about the number $s(n)$ of steps, then we can calculate:

$$s(n) = \sum_{i=1}^{n} i = \frac{n(n+1)}{2} = \frac{1}{2}n^2 + \frac{1}{2}n.$$

If we view the process pictorially, then we see the passes are tracing out a triangular half of an $n \times n$ square along the main diagonal. This intuition says "$\frac{1}{2}n^2$" without worrying about whether the main diagonal is included or excluded or "halved." The difference is a term $\frac{1}{2}n$ whose added size is relatively tiny as n becomes moderately large, so we may ignore it. Formally, we define:

DEFINITION 2.1 Two functions $s(n)$ and $t(n)$ on $\mathbb{N}$ are **asymptotically equivalent**, written $s(n) \sim t(n)$, if $\lim_{n\to\infty} \frac{s(n)}{t(n)}$ exists and equals 1.

So $s(n) \sim \frac{1}{2}n^2$, which we can also encapsulate by saying $s(n)$ is quadratic with "principal constant" $\frac{1}{2}$. But suppose now we don't know or care about the actual time units for a "step," only that the algorithm's cost scales as n^2. Another way of saying this is that as the data size n doubles, the time for the algorithm goes up by about a factor of 4. This idea doesn't care what the constant multiplying n^2 is, only that it is some constant. Hence, we define:

DEFINITION 2.2 Given two functions $s(n), t(n)$ on $\mathbb{N}$, write:

- $s(n) = O(t(n))$ if there are constants c, d such that for all n,

$$s(n) \leq c \cdot t(n) + d.$$

- $s(n) = \Omega(t(n))$ if $t(n) = O(s(n))$, and
- $s(n) = \Theta(t(n))$ if $s(n) = O(t(n))$ and $s(n) = \Omega(t(n))$.

In the first case, we say $s(n)$ is "order-of" $t(n)$ or "Big-Oh-of" $t(n)$, whereas in the second, we might say $t(n)$ is "asymptotically bounded below by" $s(n)$, and in the third, we say $s(n)$ and $t(n)$ have the same "asymptotic order."

A sufficient condition for $s(n) = \Theta(t(n))$ is that the limit $\lim_{n\to\infty} \frac{s(n)}{t(n)}$ exists and is some positive number. If the limit is zero, then we write $s(n) = o(t(n))$ instead—this "little-oh" notation is stronger than writing $s(n) = O(t(n))$ here. Thus, we can say about our $s(n)$ example above:

- $s(n) = \Theta(n^2)$;
- $s(n) = o(n^3)$;
- $\log(s(n)) = \Omega(\log n)$.

Indeed, the last gives $\log(s(n)) = \Theta(\log n)$, but it does not give $\log(s(n)) \sim \log(n)$ because the exponent 2 in $s(n)$ becomes a multiplier of 2 on the logarithm. The choice of base for the logarithm also affects the constant multiplier, but not the Θ relation. The latter enables us not to care about what the base is or even whether two logarithms have the same base. This kind of freedom is important when analyzing the costs of algorithms and even in thinking what the goals are of designing them.

One further important idea, which we will begin employing in chapter 13, uses logarithms in a different way. Write $f(n) = \tilde{O}(g(n))$ if there is some finite k such that $f(n) = O(g(n)(\log g(n))^k)$. This is pronounced "$f$ is Oh-tilde of g," and carries the idea that sometimes logarithmic as well as constant factors can be effectively ignored.

2.2 Problems

2.1. Let x be a Boolean string. What type of number does the Boolean string $x0$ represent?

2.2. Let x be a Boolean string with exactly one bit a 1. What can you say about the number it represents? Does this identification depend on using the canonical numbering of $\{0, 1\}^n$, where n is the length of x?

2.3. Compute the 4×4 "times-table" of $x \bullet y$ for $x, y \in \{00, 01, 10, 11\}$. Then write the entries in the form $(-1)^{x \bullet y}$.

2.4. Let x be a Boolean string of even length. Can the Boolean string xxx ever represent a prime number in binary notation?

2.5. Show that a function $f \colon \mathbb{N} \to \mathbb{N}$ is bounded by a constant if and only if $f(n) = O(1)$, and is linear if and only if $f(n) = \Theta(n)$.

2.6. Show that a function $f \colon \mathbb{N} \to \mathbb{N}$ is bounded by a polynomial in n, written $f(n) = n^{O(1)}$, if there is a constant C such that for all sufficiently large n, $f(2n) \leq Cf(n)$. How does C relate to the exponent k of the polynomial? Under what conditions does the converse hold? Thus, we can characterize algorithms that run in polynomial time as those for which the amount of work scales up only *linearly* as the size of the data grows linearly. Later we will use this criterion as a benchmark for *feasible* computation.

2.7. Let M be an n-bit integer, and let $a < M$. Give a bound of the form $O(s(n))$ for the time needed to find the remainder when M is divided into a^2.

2.8. Now suppose we want to compute a^{75} modulo M. Give a *concrete* bound on the number of squarings and divisions by M one needs to do, never allowing any number to become bigger than M^2.

2.9. Use the ideas of problems 2.7 and 2.8 to show that given any $a < M$, the function f_a defined for all integers $x < M$ by

$$f_a(x) = a^x \bmod M$$

can be computed in $n^{O(1)}$ time.

2.3 Summary and Notes

Numbers and strings are made out of the same "stuff," which are characters over a finite alphabet. With numbers we call them "digits," whereas with binary strings we call them "bits," but to a computer they are really the same. Switching mentally from one to the other is often a powerful way to promote one's understanding of theoretical concepts. This is especially true in quantum computations, where the bits of a Boolean string—or rather their indexed locations in the string—will be treated as *quantum coordinates* or *qubits*. The next chapter shows how we use both numbers and strings as indices to vectors and matrices.

It is interesting that while the notion of natural numbers is ancient, the notion of Boolean strings is much more recent. Even more interesting is that it is only with the rise of computing that the importance of using just the two Boolean values $0, 1$ has become so clear. Asymptotic notation helped convince us that whether we operate in base 10 or 2 or 16 or 64, the difference is secondary compared to the top-level structure of the algorithm, which usually determines the asymptotic order of the running time.

3 Basic Linear Algebra

A vector $\boldsymbol{a}$ of dimension N is an ordered list of values, just as usual. Thus, $\boldsymbol{a}$ of dimension N stands for:

$$\begin{bmatrix} a_0 \\ a_1 \\ \vdots \\ a_{N-1}. \end{bmatrix}$$

Instead of the standard subscript notation to denote the k-th element, we will use $\boldsymbol{a}(k)$ to denote it. This has the advantage of making some of our equations a bit more readable, but just remember that

$$\boldsymbol{a}(k) \text{ is the same as } a_k.$$

One rationale for this notation is that a vector can often be best viewed as being indexed by other sets rather than just the usual $0,\ldots,N-1$. The functional notation $\boldsymbol{a}(k)$ seems to be more flexible in this regard.

A fancier justification is that sometimes vector spaces are defined as functions over sets, so this notation is consistent with that. In any event, $\boldsymbol{a}(k)$ is just the element of the vector that is indexed by k. A concrete example is to consider a vector $\boldsymbol{a}$ of dimension 4. In this case, we may use the notation

$$\boldsymbol{a}(0),\boldsymbol{a}(1),\boldsymbol{a}(2),\boldsymbol{a}(3),$$

for its elements, or we may employ the notation

$$\boldsymbol{a}(00),\boldsymbol{a}(01),\boldsymbol{a}(10),\boldsymbol{a}(11),$$

using Boolean strings to index the elements.

A philosophical justification for our functional notation is that vectors are pieces of code. We believe that nature computes with code—not with the graph of the code. For instance, each elementary standard basis vector $\boldsymbol{e}_k$ is 0 except for the k-th coordinate, which is 1, and we use the subscript when thinking of it as an object. When $N = 2^n$, we index the complex coordinates from 0 as $0,\ldots,N-1$ and enumerate $\{0,1\}^n$ as $x_0,\ldots,x_{N-1}$, but we index the binary string places from 1 as $1,\ldots,n$. Doing so helps tell them apart. When thinking of $\boldsymbol{e}_k$ as a piece of code, we get the function $\boldsymbol{e}_k(x) = 1$ if $x = x_k$ and $\boldsymbol{e}_k(x) = 0$ otherwise. We can also replace k by a binary string as a subscript, for instance, writing the four standard basis vectors when $n = 2$ as $\boldsymbol{e}_{00},\boldsymbol{e}_{01},\boldsymbol{e}_{10},\boldsymbol{e}_{11}$.

3.1 Hilbert Spaces

A real Hilbert space is nothing more than a fancy name for the usual Euclidean space. We will use $\mathbb{H}_N$ to denote this space of dimension N. The elements are real vectors $\boldsymbol{a}$ of dimension N. They are added and multiplied by scalars in the usual manner:

- If $\boldsymbol{a}, \boldsymbol{b}$ are vectors in this space, then so is $\boldsymbol{a}+\boldsymbol{b}$, which is defined by

$$\begin{bmatrix} \boldsymbol{a}(0) \\ \vdots \\ \boldsymbol{a}(N-1) \end{bmatrix} + \begin{bmatrix} \boldsymbol{b}(0) \\ \vdots \\ \boldsymbol{b}(N-1) \end{bmatrix} = \begin{bmatrix} \boldsymbol{a}(0)+\boldsymbol{b}(0) \\ \vdots \\ \boldsymbol{a}(N-1)+\boldsymbol{b}(N-1) \end{bmatrix}.$$

- If $\boldsymbol{a}$ is a vector again in this space and c is a real number, then $\boldsymbol{b} = c\boldsymbol{a}$ is defined by

$$c \begin{bmatrix} \boldsymbol{a}(0) \\ \vdots \\ \boldsymbol{a}(N-1) \end{bmatrix} = \begin{bmatrix} c\boldsymbol{a}(0) \\ \vdots \\ c\boldsymbol{a}(N-1) \end{bmatrix}.$$

The abstract essence of a Hilbert space is that each vector has a **norm**: The norm of a vector $\boldsymbol{a}$, really just its length, is defined to be

$$||\boldsymbol{a}|| = \left| \sum_k \boldsymbol{a}(k)^2 \right|^{1/2}.$$

Note that in the case of two dimensions, the norm of the vector

$$\boldsymbol{a} = \begin{bmatrix} r \\ s \end{bmatrix}$$

is its usual length in the plane, $\sqrt{r^2+s^2}$. A Hilbert space simply generalizes this to many dimensions. A **unit vector** is just a vector of norm 1. Unit vectors together comprise the **unit sphere** in any Hilbert space.

3.2 Products and Tensor Products

The ordinary Cartesian product of an m-dimensional Hilbert space $\mathbb{H}_1$ and an n-dimensional Hilbert space $\mathbb{H}_2$ is the $(m+n)$-dimensional Hilbert space obtained by concatenating vectors from the former with vectors from the latter. Its vectors have the form $\boldsymbol{a}(i)$ with $i \in [m+n]$.

Their *tensor product* $\mathbb{H}_1 \otimes \mathbb{H}_2$, however, has vectors of the form $\boldsymbol{a}(k)$, where $1 \leq k \leq mn$. Indeed, k is in 1–1 correspondence with pairs (i,j) of indices where $i \in [m]$ and $j \in [n]$. Because we regard indices as strings, we can write them juxtaposed as $\boldsymbol{a}(ij)$.

The tensor product of two vectors $\boldsymbol{a}$ and $\boldsymbol{b}$ is the vector $\boldsymbol{c} = \boldsymbol{a} \otimes \boldsymbol{b}$ defined by

$$\boldsymbol{c}(ij) = \boldsymbol{a}(i)\boldsymbol{b}(j).$$

A vector denoting a pure quantum state is **separable** if it is the tensor product of two other vectors; otherwise it is **entangled**. The vectors $\boldsymbol{e}_{00}$ and $\boldsymbol{e}_{11}$ are separable, but their unit-scaled sum $\frac{1}{\sqrt{2}}(\boldsymbol{e}_{00} + \boldsymbol{e}_{11})$ is entangled. The standard basis vectors of $\mathbb{H}_1 \otimes \mathbb{H}_2$ are separable, but this is not true of many of their linear combinations, all of which still belong to $\mathbb{H}_1 \otimes \mathbb{H}_2$ because it is a Hilbert space.

Often our quantum algorithms will operate on the product of two Hilbert spaces, each using binary strings as indices, which gives us the space of vectors of the form $\boldsymbol{a}(xy)$. Here x ranges over the indices of the first space and y over those of the second space. Writing $\boldsymbol{a}(xy)$ does not entail that $\boldsymbol{a}$ is separable.

3.3 Matrices

Matrices represent linear operators on Hilbert spaces. We can add them together, we can multiply them, and of course we can use them to operate on vectors. We assume these notions are familiar to you; if not, please see sources in this chapter's end notes. A typical example is

$$\boldsymbol{UVa} = \boldsymbol{b}.$$

This means: apply the $\boldsymbol{V}$ transform to the vector $\boldsymbol{a}$, then apply the $\boldsymbol{U}$ transform to the resulting vector, and the answer is the vector $\boldsymbol{b}$. The matrix $\boldsymbol{I}_N$ denotes the $N \times N$ identity matrix. We use square brackets for matrix entries to distinguish them further from vector amplitudes, so that the identity matrix has entries $\boldsymbol{I}_N[r,c] = 1$ if $r = c$, $\boldsymbol{I}_N[r,c] = 0$ otherwise. One of the key properties of matrices is that they define linear operations, namely:

$$\boldsymbol{U}(\boldsymbol{a} + \boldsymbol{b}) = \boldsymbol{Ua} + \boldsymbol{Ub}.$$

The fact that all our transformations are linear is what makes quantum algorithms so different from classical ones. This will be clearer as we give examples, but the linearity restriction both gives quantum algorithms great power and curiously makes them so different from classical ones.

If $\boldsymbol{U}$ is a matrix, then we use $\boldsymbol{U}^k$ to denote the k-th power of the matrix,

$$\boldsymbol{U}^k = \underbrace{\boldsymbol{U}\boldsymbol{U}\cdots\boldsymbol{U}}_{k \text{ copies}}.$$

DEFINITION 3.1 The **transpose** of a matrix $\boldsymbol{U}$ is the matrix $\boldsymbol{V}$ such that

$$\boldsymbol{V}[r,c] = \boldsymbol{U}[c,r].$$

We use $\boldsymbol{U}^T$ as usual to denote the transpose of a matrix. We also use transpose for vectors but only when writing them after a matrix in lines of text. Generally, we minimize the distinction between row and column vectors, using the latter as standard. The **inner product** of two *real* vectors $\boldsymbol{a}$ and $\boldsymbol{b}$ is given by

$$\langle \boldsymbol{a}, \boldsymbol{b} \rangle = \sum_{k=0}^{m} \boldsymbol{a}(k)\boldsymbol{b}(k).$$

DEFINITION 3.2 A real matrix $\boldsymbol{U}$ is **unitary** provided $\boldsymbol{U}^T\boldsymbol{U} = \boldsymbol{I}$.

Here are three unitary 2×2 real matrices. Note that the last, called the **Hadamard matrix**, requires a constant multiplier to divide out a factor of 2 that comes from squaring it.

$$\boldsymbol{I} = \begin{bmatrix} 1 & 0 \\ 0 & 1 \end{bmatrix}, \quad \boldsymbol{X} = \begin{bmatrix} 0 & 1 \\ 1 & 0 \end{bmatrix}, \quad \boldsymbol{H} = \frac{1}{\sqrt{2}} \begin{bmatrix} 1 & 1 \\ 1 & -1 \end{bmatrix}.$$

The first two are also **permutation matrices**, meaning square matrices each of whose rows and columns has all zeros except for a single 1. All permutation matrices are unitary.

Another definition of a unitary matrix is based on the notion of orthogonality. Call two vectors $\boldsymbol{a}$ and $\boldsymbol{b}$ **orthogonal** if their inner product is 0. A matrix $\boldsymbol{U}$ is unitary proved each row is a unit vector and any two distinct rows are orthogonal. The reason these matrices are so important is that they preserve the Euclidean length.

LEMMA 3.3 If $\boldsymbol{U}$ is a unitary matrix and $\boldsymbol{a}$ is a vector, then $||\boldsymbol{U}\boldsymbol{a}|| = ||\boldsymbol{a}||$.

Proof. Direct calculation gives:

$$\begin{aligned}
||\boldsymbol{Ua}||^2 &= \sum_x |\boldsymbol{Ua}(x)|^2 \\
&= \sum_x \boldsymbol{Ua}(x) \cdot \boldsymbol{Ua}(x) \\
&= \sum_x (\sum_y \boldsymbol{U}[x,y]\boldsymbol{a}(y))(\sum_z \boldsymbol{U}[x,z]\boldsymbol{a}(z)) \\
&= \sum_x (\sum_y \sum_z \boldsymbol{U}[x,y]\boldsymbol{U}[x,z]\boldsymbol{a}(y)\boldsymbol{a}(z)) \\
&= \sum_y \sum_z (\sum_x (\boldsymbol{U}[x,y]\boldsymbol{U}[x,z]))\boldsymbol{a}(y)\boldsymbol{a}(z)) \\
&= \sum_y \boldsymbol{a}(y)\boldsymbol{a}(y) = ||\boldsymbol{a}||^2
\end{aligned}$$

because the inner product of $\boldsymbol{U}[-,y]$ and $\boldsymbol{U}[-,z]$ is 1 or 0 according as $y = z$.
□

As far as we can, we try not to care about whether a Hilbert space is real or complex. However, we do need notation for complex spaces.

3.4 Complex Spaces and Inner Products

To describe some algorithms, mainly Shor's, we need to use complex vectors and matrices. In this case, all definitions are the same except for the notion of transpose and inner product. Both now need to use the conjugation operation: if $z = x + iy$ where x, y are real numbers and $i = \sqrt{-1}$ as usual, then the **conjugate** of z is $x - iy$ and is denoted by $\overline{z}$. We now define the **adjoint** of a matrix $\boldsymbol{U}$ to be the matrix $\boldsymbol{V} = \boldsymbol{U}^*$ such that

$$\boldsymbol{V}[r,c] = \overline{\boldsymbol{U}}[c,r].$$

A complex matrix $\boldsymbol{U}$ is **unitary** provided $\boldsymbol{U}^*\boldsymbol{U} = \boldsymbol{I}$. Furthermore, the inner product of two complex vectors $\boldsymbol{a}$ and $\boldsymbol{b}$ is defined to be

$$\langle \boldsymbol{a}, \boldsymbol{b} \rangle = \sum_{k=0}^{m} \overline{\boldsymbol{a}(k)}\boldsymbol{b}(k).$$

Note that because $\overline{r}$ is the same as r for a real number r, these concepts are the same as what we defined before when the entries of the vectors and matrices are all real numbers. Rather then use special notation for the complex case, we will use the same as in the real case—the context should make it clear which is being used. The one caveat is that in the few cases where the entries of $\boldsymbol{a}$ or $\boldsymbol{U}$ are complex, one needs to conjugate them. Upon observing this, the proof of the following is much the same as for Lemma 3.3 above:

LEMMA 3.4 If $\boldsymbol{U}$ is a unitary matrix and $\boldsymbol{a}$ is a vector, then the length of $\boldsymbol{Ua}$ is the same as the length of $\boldsymbol{a}$.

We can also form tensor products of matrices having any dimensions. If $\boldsymbol{U}$ is $m \times n$ and $\boldsymbol{V}$ is $r \times s$, then $\boldsymbol{W} = \boldsymbol{U} \otimes \boldsymbol{V}$ is the $mr \times ns$ matrix whose action on *product* vectors $\boldsymbol{c}(ij) = \boldsymbol{a}(i)\boldsymbol{b}(j)$ is as follows:

$$(\boldsymbol{Wc})(ij) = (\boldsymbol{Ua})(i)(\boldsymbol{Vb})(j).$$

Because every vector $\boldsymbol{d}$ of dimension rs (whether entangled or not) can be written as a linear combination of basis vectors, each of which is a product vector, the action $\boldsymbol{Wd}$ is well defined via the same linear combination of the outputs on the basis vectors. That is, if

$$\boldsymbol{d} = \sum_{i=1}^{r} \sum_{j=1}^{s} d_{i,j} \boldsymbol{e}_i \otimes \boldsymbol{e}_j,$$

then

$$\boldsymbol{Wd} = \sum_{i=1}^{r} \sum_{j=1}^{s} d_{i,j} (\boldsymbol{Ue}_i) \otimes \left(\boldsymbol{Ve}_j\right).$$

Note that it does not matter whether the scalars $d_{i,j}$ are regarded as multiplying $\boldsymbol{e}_i$, $\boldsymbol{e}_j$, or the whole thing. This fact matters later in section 6.5. We mainly use tensor products to combine operations that work on separate halves of an overall index ij.

3.5 Matrices, Graphs, and Sums Over Paths

One rich source of real-valued matrices is *graphs*. A **graph** G consists of a set V of **vertices**, also called **nodes**, together with a binary relation E on V whose members are called **edges**. The **adjacency matrix** $\boldsymbol{A} = \boldsymbol{A}(G)$ of a graph $G = (V, E)$ is defined for all $u, v \in V$ by:

$$\boldsymbol{A}[u, v] = \begin{cases} 1 & \text{if } (u, v) \in E \\ 0 & \text{otherwise.} \end{cases}$$

If E is a symmetric relation, then $\boldsymbol{A}$ is a symmetric matrix. In that case, G is an **undirected graph**; otherwise it is **directed**.

The **degree** of a vertex u is the number of edges incident to u, which is the same as the number of 1s in row u of $\boldsymbol{A}$. A graph is **regular** of degree d if every vertex has degree d. In that case, consider the matrix $\boldsymbol{A}' = \frac{1}{d}\boldsymbol{A}$. Figure 3.1 exemplifies this for a graph called the four-cycle, C_4. This graph is **bipartite**, meaning that V can be partitioned into V_1, V_2 such that every edge connects a vertex in V_1 and a vertex in V_2.

Figure 3.1
Four-cycle graph $G = C_4$, stochastic adjacency matrix $\boldsymbol{A}'_G$, and unitary matrix $\boldsymbol{U}_G$.

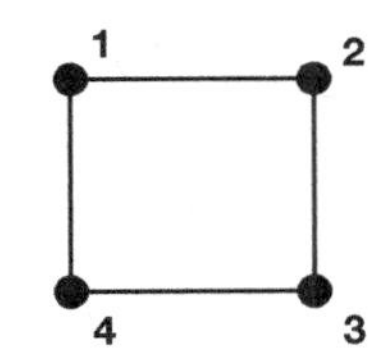

$$\boldsymbol{A}'_G = \frac{1}{2}\begin{bmatrix} 0 & 1 & 0 & 1 \\ 1 & 0 & 1 & 0 \\ 0 & 1 & 0 & 1 \\ 1 & 0 & 1 & 0 \end{bmatrix}, \qquad \boldsymbol{U}_G = \frac{1}{\sqrt{2}}\begin{bmatrix} 0 & 1 & 0 & 1 \\ 1 & 0 & 1 & 0 \\ 0 & 1 & 0 & -1 \\ 1 & 0 & -1 & 0 \end{bmatrix}.$$

Because every row has non-negative numbers summing to 1, $\boldsymbol{A}'$ is a **stochastic matrix**. Because the columns also sum to 1, $\boldsymbol{A}'$ is **doubly stochastic**.

However, $\boldsymbol{A}'$ is not unitary for two reasons. First, the Euclidean norm of each row and column is $(\frac{1}{2})^2(1+1) = \frac{1}{2}$, not 1. Second, not all pairs of distinct rows or columns are orthogonal. To meet the first criterion, we multiply by $\frac{1}{\sqrt{2}}$ instead of $\frac{1}{2}$. To meet the second, we can change the entries for the edge between nodes 3 and 4 from 1 to -1, creating the matrix $\boldsymbol{U}_G$, which is also shown in figure 3.1. Then $\boldsymbol{U}_G$ is unitary—in fact, it is the tensor product $\boldsymbol{H} \otimes \boldsymbol{X}$ of two 2×2 unitary matrices given after definition 3.2.

This fact is peculiar to the four-cycle. An example of a d-regular graph whose adjacency matrix cannot similarly be converted into a unitary matrix is the 3-regular prism graph shown in figure 3.2. Rows 1 and 3 have a nonzero dot product in a single column, namely, column 5, so there is no substitution of nonzero values for nonzero entries that will make them orthogonal.

Figure 3.2
3-regular prism graph G and stochastic adjacency matrix $\boldsymbol{A}'$.

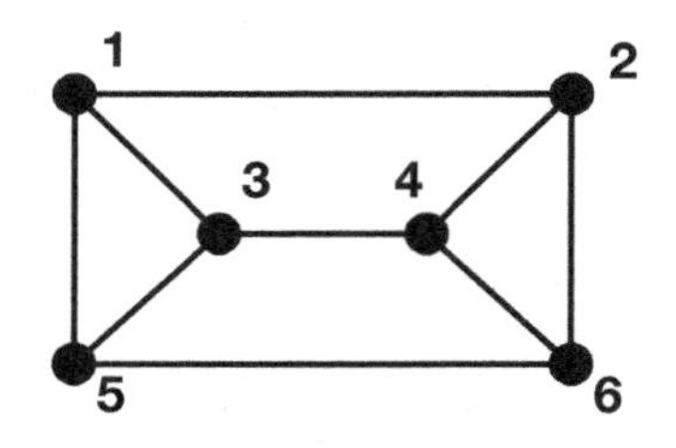

$$\boldsymbol{A}' = \frac{1}{3}\begin{bmatrix} 0 & 1 & 1 & 0 & 1 & 0 \\ 1 & 0 & 0 & 1 & 0 & 1 \\ 1 & 0 & 0 & 1 & 1 & 0 \\ 0 & 1 & 1 & 0 & 0 & 1 \\ 1 & 0 & 1 & 0 & 0 & 1 \\ 0 & 1 & 0 & 1 & 1 & 0 \end{bmatrix}$$

However, in chapter 14, we will see a general technique involving a tensor product of $\boldsymbol{A}_G$ with another matrix to create a unitary matrix, one representing a quantum rather than a classical random walk on the graph G.

The square of an adjacency matrix has entries given by

$$\boldsymbol{A}^2[i,j] = \sum_k \boldsymbol{A}[i,k]\boldsymbol{A}[k,j].$$

The sum counts the number of ways to go from vertex i through some vertex k and end up at vertex j. That is, $\boldsymbol{A}^2[i,j]$ counts the number of *paths* of *length* 2 from i to j. Likewise, $\boldsymbol{A}^3[i,j]$ counts paths of length exactly 3, $\boldsymbol{A}^4[i,j]$ those of length 4, and so on.

For directed graphs in which each edge goes from one of n *sources* to one of m *sinks*, the adjacency matrices $\boldsymbol{A}_{n,m}$ lend themselves to a path-counting composition. Consider

$$\boldsymbol{U} = \boldsymbol{A}_{n,m_1}\boldsymbol{A}_{m_1,m_2}\boldsymbol{A}_{m_2,m_3}\cdots\boldsymbol{A}_{m_k,r}.$$

Note how the m_i dimensions match like dominoes. This makes $\boldsymbol{U}$ a legal matrix product. The corresponding graph identifies the sinks of the graph G_ℓ represented by $\boldsymbol{A}_{m_{\ell-1},m_\ell}$ (here $m_0 = n$) with the sources of $G_{\ell+1}$. Then $\boldsymbol{U}[i,j]$ counts the number of paths in the product graph that begin at source node i of G_1 and end at sink node j of the last graph G_{k+1}.

What links quantum and classical concepts here is that *weights* can be put on the edges of these graphs, even complex-number weights. When these are substituted for the "1" entries of the adjacency matrices, the value $\boldsymbol{U}[i,j]$ becomes a sum of products over the possible paths. In the classical case, these weights can be probabilities of individual choices along the paths. In the quantum case, they can be complex *amplitudes*, and some of the products in the sum may *cancel*. Either way, the leading algorithmic structure is that of a *sum over paths*. As a fundamental idea in quantum mechanics, it was advanced particularly by Richard Feynman, and yet it needs no more than this much about graphs and linear algebra to appreciate.

This discussion already conveys the basic flavor of quantum operations. The matrices are unitary rather than stochastic; the entries are square-roots of probabilities rather than probabilities; negative entries—and later imaginary number entries—are used to achieve cancellations. After defining feasible computations, we will examine some important unitary matrices in greater detail.

3.6 Problems

3.1. Show that the product of unitary matrices is unitary.

3.2. If $\boldsymbol{U}$ is a matrix, then what is $\boldsymbol{U}\boldsymbol{e}_k$?

3.3. Show that the *columns* of a unitary matrix are unit vectors. Also show that distinct columns are orthogonal.

3.4. Consider the matrix

$$\begin{bmatrix} w & w \\ w & -w \end{bmatrix}.$$

For what real values of w is it a unitary matrix?

3.5. Consider the matrix $\boldsymbol{U}$ equal to:

$$\frac{1}{\sqrt{2}}\begin{bmatrix} \boldsymbol{I}_N & \boldsymbol{I}_N \\ \boldsymbol{I}_N & -\boldsymbol{I}_N \end{bmatrix}.$$

Show that for any N, the matrix $\boldsymbol{U}$ is unitary.

3.6. For real vectors $\boldsymbol{a}$ and $\boldsymbol{b}$, show that the inner product can be defined from the norm as:

$$(\boldsymbol{a},\boldsymbol{b}) = 1/2\left(||\boldsymbol{a}+\boldsymbol{b}||^2 - ||\boldsymbol{a}||^2 - ||\boldsymbol{b}||^2\right).$$

3.7. For any complex $N \times N$ matrix $\boldsymbol{U}$, we can uniquely write $\boldsymbol{U} = \boldsymbol{R} + i\boldsymbol{Q}$, where $\boldsymbol{Q}$ and $\boldsymbol{R}$ have real entries. Show that if $\boldsymbol{U}$ is unitary, then so is the $2N \times 2N$ matrix $\boldsymbol{U}'$ given in block form by

$$\boldsymbol{U}' = \begin{bmatrix} \boldsymbol{R} & \boldsymbol{Q} \\ -\boldsymbol{Q} & \boldsymbol{R} \end{bmatrix}.$$

Thus, by doubling the dimension, we can remove the need for complex-number entries.

3.8. Apply the construction of the last problem to the matrix

$$\boldsymbol{Y} = \begin{bmatrix} 0 & -i \\ i & 0 \end{bmatrix}.$$

This is the second of the so-called **Pauli matrices**, along with $\boldsymbol{X}$ above and $\boldsymbol{Z}$ defined below.

3.9. Consider the following matrix:

$$\boldsymbol{V} = \frac{1}{\sqrt{2}}\begin{bmatrix} e^{i\pi/4} & e^{-i\pi/4} \\ e^{-i\pi/4} & e^{i\pi/4} \end{bmatrix} = \frac{1}{2}\begin{bmatrix} 1+i & 1-i \\ 1-i & 1+i \end{bmatrix}.$$

What is $\boldsymbol{V}^2$?

3.10. Let $\boldsymbol{T}_\alpha$ denote the 2×2 "twist" matrix

$$\begin{bmatrix} 1 & 0 \\ 0 & e^{i\alpha} \end{bmatrix},$$

respectively. Show that it is unitary. Also find a complex scalar c such that $c\boldsymbol{T}_\alpha$ has determinant 1 and write out the resulting matrix, which often also goes under the name $\boldsymbol{T}_\alpha$.

3.11. The following cases of $\boldsymbol{T}_\alpha$ for $\alpha = \pi, \frac{\pi}{2}, \frac{\pi}{4}$ have special names as shown:

$$\boldsymbol{Z} = \begin{bmatrix} 1 & 0 \\ 0 & -1 \end{bmatrix}, \quad \boldsymbol{S} = \begin{bmatrix} 1 & 0 \\ 0 & i \end{bmatrix}, \quad \boldsymbol{T} = \begin{bmatrix} 1 & 0 \\ 0 & e^{i\pi/4} \end{bmatrix}.$$

How are they related? Also find an equation relating the Pauli matrices $\boldsymbol{X}, \boldsymbol{Y}, \boldsymbol{Z}$ along with an appropriate "phase scalar" c as in the last problem.

3.12. In this and the next problem, consider the following *group commutators* of the three matrices of the last problem with the Hadamard matrix $\boldsymbol{H}$ (noting also that $\boldsymbol{H}^* = \boldsymbol{H}^{-1} = \boldsymbol{H}$, i.e., the Hadamard is **self-adjoint** as a unitary matrix):

$$\begin{aligned} \boldsymbol{Z}' &= \boldsymbol{HZHZ}^* \\ \boldsymbol{S}' &= \boldsymbol{HSHS}^* \\ \boldsymbol{T}' &= \boldsymbol{HTHT}^* \end{aligned}$$

Show that $\boldsymbol{Z}'$ and some multiple $c\boldsymbol{S}'$ have nonzero entries that are powers of i.

3.13. Show, however, that no multiple $c\boldsymbol{T}'$ has entries of this form by considering the mutual angles of its entries. What is the lowest power 2^r such that these angles are multiples of $\pi/2^r$?

3.14. Define a matrix to be *balanced* if all of its nonzero entries have the same magnitude. Of all the 2×2 matrices in problem 3.12, say which are balanced. Is the property of being balanced closed under multiplication?

3.15. Show that the **rotation matrix** by an angle θ (in the real plane),

$$\boldsymbol{R}_x(\theta) = \begin{bmatrix} \cos(\theta/2) & \sin(\theta/2) \\ -\sin(\theta/2) & \cos(\theta/2) \end{bmatrix}$$

is unitary. Also, what does $\boldsymbol{R}_x^2$ represent?

3.16. Show that for every 2×2 unitary matrix $\boldsymbol{U}$, there are real numbers $\theta, \alpha, \beta, \delta$ such that

$$\boldsymbol{U} = e^{i\delta}\boldsymbol{T}_\alpha \boldsymbol{R}_\theta \boldsymbol{T}_\beta.$$

Thus, every 2×2 unitary operation can be decomposed into a rotation flanked by two twists, multiplied by an arbitrary phase shift by δ. Write out the decomposition for the matrix $\boldsymbol{V}$ in problem 3.9. (It doesn't matter which definition of "$\boldsymbol{T}_\alpha$" you use from problem 3.10.)

3.17. Show how to write $\boldsymbol{V}$ as a composition of Hadamard and $\boldsymbol{T}$ matrices. (All of these problems point out the special nature of the $\boldsymbol{T}$-matrix. $\boldsymbol{T}$ is often called the "$\pi/8$ gate"; the confusing difference from $\pi/4$ owes to the constant c in problem 3.10.)

3.18. Show that the four Pauli matrices, $\boldsymbol{I}$, $\boldsymbol{X}$, $\boldsymbol{Y}$, and $\boldsymbol{Z}$, form an orthonormal basis for the space of 2×2 matrices, regarded as a 4-dimensional complex Hilbert space.

3.19. Define G to be the complete undirected graph on 4 vertices, whose 6 edges connect every pair (i,j) with $i \neq j$. Convert its adjacency matrix $\boldsymbol{A}_G$ into a unitary matrix by making some entries negative and multiplying by an appropriate constant. Can you do this in a way that preserves the symmetry on the matrix?

3.20. Show that a simple undirected graph G has a triangle if and only if $\boldsymbol{A}^2 \odot \boldsymbol{A}$ has a nonzero entry, where $\boldsymbol{A} = \boldsymbol{A}_G$ and $\odot$ here means multiplying entry-wise. This means that triangle-detection in an n-vertex graph has complexity no higher than that of squaring a matrix, which is equivalent to that of multiplying two $n \times n$ matrices.

Surprisingly, the obvious matrix multiplication algorithm and its $O(n^3)$ running time are far from optimal, and the current best exponent on the "n" for matrix multiplication is about 2.372. This certainly improves on the $O(n^3)$ running time of the simple algorithm that tries every set of three vertices to see if it forms a triangle.

3.7 Summary and Notes

There are many good linear algebra texts and online materials. If this material is new or you took a class on it once and need a refresher, here are some suggested places to go:

- The textbook *Elementary Linear Algebra* by Kuttler (2012).
- Linear algebra video lectures by Gilbert Strang which are maintained at MITOPENCOURSEWARE: http://ocw.mit.edu/courses/mathematics/18-06-linear-algebra-spring-2010/video-lectures/
- The textbook *Graph Algorithms in the Language of Linear Algebra* by Kepner and Gilbert (2011).

The famous text by Nielsen and Chuang (2000) includes a full treatment of linear algebra needed for quantum computation. Feynman (1982, 1985) wrote the first two classic papers on quantum computation.

4 Boolean Functions, Quantum Bits, and Feasibility

A *Boolean function* f is a mapping from $\{0,1\}^n$ to $\{0,1\}^m$, for some numbers n and m. When we define a Boolean function $f(x_1,\dots,x_n) = (y_1,\dots,y_m)$, we think of the x_i as inputs and the y_j as outputs. We also regard the x_i together as a binary string x and similarly write y for the output. When $m = 1$, there is some ambiguity between the output as a string or a single bit because we write just "y" not "(y)" in the latter case as well, but the difference does not matter in context. When $m = 1$, you can also think of f as a predicate: x *satisfies* the predicate if and only if $f(x) = 1$.

Thus, Boolean functions give us all of the following: the basic truth values, binary strings, and, as seen in chapter 2, also numbers and other objects. The most basic have $n = 1$ or 2, such as the unary NOT function, and binary AND, OR, and XOR. We can also regard the following higher-arity versions as basic:

- AND: This is the function $f(x_1,\dots,x_n)$ defined as 1 if and only if every argument is 1. Thus,

$$f(1,1,1) = 1 \text{ and } f(1,0,1,1) = 0.$$

- OR: This is the function $f(x_1,\dots,x_n)$ defined as 1 if the number of 1's in $x_1,\dots,x_n$ is non-zero. Thus,

$$f(0,1,1) = 1 \text{ and } f(0,0,0,0) = 0.$$

- XOR: This is the function $f(x_1,\dots,x_n)$ defined as 1 if the number of 1's in $x_1,\dots,x_n$ is odd. Thus,

$$f(0,1,1) = 0 \text{ and } f(1,1,1,1,1) = 1.$$

  The latter is true because there are five 1's.

The binary operations can also be applied on pairs of strings bitwise. For instance, if x and y are both Boolean strings of length n, then $x \oplus y$ is equal to

$$z = (x_1 \oplus y_1,\dots,x_n \oplus y_n).$$

We could similarly define the bitwise-AND and the bitwise-OR of two equal-length binary strings. These are not the same as the above n-ary operations but are instead n applications of binary operations. Each operation connects the i-th bit of x with then i-th bit of y, for some i, and they intuitively run "in parallel." The Boolean inner product, which we defined in chapter 2, is computed by feeding the bitwise binary AND into the n-ary XOR, that is:

$$x \bullet y = \mathsf{XOR}(x_1 \wedge y_1,\dots,x_n \wedge y_n).$$

In cases like $x \oplus y$, we say we have a *circuit* of n *Boolean gates* that collectively compute the Boolean function $f\colon \{0,1\}^r \to \{0,1\}^n$, with $r = 2n$, defined by $f(x,y) = x \oplus y$. Technically, we need to specify whether x and y are given sequentially as $(x_1,\ldots,x_n,y_1,\ldots,y_n)$ or shuffled as $(x_1,y_1,\ldots,x_n,y_n)$, and this matters to how we would draw the circuit as a picture. But either way we have a $2n$-input function that represents the same function of two n-bit strings.

The number of gates is identified with the amount of *work* or *effort* expended by the circuit, and this in turn is regarded as the sequential *time* for the circuit to execute. It does not matter too much whether one counts gates or *wires* between gates. What is critical is that only basic operations can be used, and that they can only apply to previously computed values. In the following sketch, the NOT of $a \vee b$ is allowed because $a \vee b$ has already been computed:

$$\ldots(a \vee b)\ldots\ldots\neg(a \vee b)\ldots$$

Two Boolean functions that we should *not* regard as basic are:

- PRIME: This is the function $f(x_1,\ldots,x_n)$ defined as 1 if the Boolean string $x = x_1,\ldots,x_n$ represents a number that is a prime number. Recall a prime number is a natural number p greater than 1 with only 1 and p as divisors.
- FACTOR: This is the function $f(x_1,\ldots,x_n,w_1,\ldots,w_n)$ regarded as having two integers x and w as arguments—note that we can pad w as well as x by leading 0's. It returns 1 if and only if x has no divisor greater than w, aside from x itself.

The game is, how efficiently can we build a circuit to compute these functions? They are related by $\mathsf{PRIME}(x) = \mathsf{FACTOR}(x,1)$ for all x. This implies that a circuit for FACTOR immediately gives one for solving the predicate PRIME because one can simply fix the "w" inputs to be the padded version of the number 1. This does not imply the converse relation, however. Although both of these functions have been studied for 3,000 years, PRIME was shown only a dozen years ago to be *feasible* in a sense we describe next, while many believe that FACTOR is not feasible at all. Unless you are allowed a *quantum* circuit, that is.

4.1 Feasible Boolean Functions

Not all Boolean functions are created equal; some are more complex than others. In the above examples, which n-bit function would you like to compute if you had to? I think we all will agree that the easiest is the OR function: just glance at the input bits and check whether one is a 1. If you only see 0's, then clearly the OR function is 0. AND is similar.

Next harder it would seem is the XOR function. The intuitive reason is that now you have to count the number of bits, and this count has to be exact. If there are 45 inputs that are 1 and you miscount and think there are 44, then you will get the wrong value for the function. Indeed, one can argue that n-ary XOR is harder than the bitwise-XOR function because each of the n binary XOR operations is "local" on its own pair of bits.

More difficult is the PRIME function. There is no known algorithm that we can use and just glance at the bits. Is

$$101010110110101101011110101$$

a prime number or not? Of course you first might convert it to a decimal number: it represents the number 11234143. This still now requires some work to see if it has any nontrivial divisors, but it does: 23 and 488441.

One of the achievements of computer science is that we can define the classical complexity of a Boolean function. Thus, AND and XOR are computable in a linear number of steps, that is, $O(n)$. Known circuits for PRIME take more than linearly many steps, but the time is still *polynomial*, that is, $n^{O(1)}$. But there are also Boolean functions that require time exponential in n. Many people believe that FACTOR is one of them, but nobody knows for sure.

To see the issue, consider that any Boolean function can be defined by its *truth table*. Here is the truth table for the exclusive-or function XOR:

x	y	$x \oplus y$
0	0	0
0	1	1
1	0	1
1	1	0

Each row of the table tells you what the function, in this case, the exclusive-or function, does on the inputs of that row. In general, a Boolean function $f(x_1,\ldots,x_n)$ is defined by a truth table that has 2^n rows—one for each possible input. Thus, if $n = 3$, there are eight possible rows:

$$000, 001, 010, 011, 100, 101, 110, \text{ and } 111.$$

The difficulty is that as the number of inputs grows, the truth table increases exponentially in size.

Thus, representing Boolean functions by their truth tables is alway *possible*, but is not always *feasible*. The tables will be large when there are thirty inputs, and when there are over 100, the table would be impossible to write down.

The final technical concept we need is having not just a single Boolean function but rather a *family* $[f_n]$ of Boolean functions, each f_n taking n inputs, that are conceptually related. That is, the $[f_n]$ constitute a single function f on strings of all lengths, so we write $f\colon \{0,1\}^* \to \{0,1\}$, or for general rather than one-bit outputs, $f\colon \{0,1\}^* \to \{0,1\}^*$. Maybe it is confusing to write "f" also for this kind of function with an infinite domain, but the intent is usually transparent—as when letting AND, OR, and XOR above apply to any n. Now we can finally define "feasible":

DEFINITION 4.1 A Boolean function $f = [f_n]$ is **feasible** provided the individual f_n are computed by circuits of size $n^{O(1)}$.

4.2 An Example

Consider the Boolean function $\text{MAJ}(x_1,x_2,x_3,x_4,x_5)$, which takes the majority of five Boolean inputs. A first idea is to compute it using applications of OR and AND as follows: For every three-element subset $S = \{i,j,k\}$ of $\{1,2,3,4,5\}$, we compute $y_S = \text{OR}(x_i,x_j,x_k)$. Define y to be the AND of each of the ten subsets S. Then $y = 1 \iff$ no more than 2 bits of $x_1,\dots,x_5$ are 0 $\iff \text{MAJ}(x_1,x_2,x_3,x_4,x_5)$ is true. The complexity is counted as 11 operations and, importantly, $10 \times 3 + 10 = 40$ total arguments of those operations.

On second thought, we can find a program of slightly lower complexity. Consider the Boolean circuit diagram in figure 4.1.

Expressed as a sequence of operations, in one of many possible orders, the circuit is equivalent to the following *straight-line program*:

$$\begin{aligned} v_1 &= \text{OR}(x_1,x_2,x_3), \quad v_2 = \text{OR}(x_4,x_5), \\ w_1 &= \text{AND}(x_1,x_2), \quad w_2 = \text{AND}(x_1,x_3), \quad w_3 = \text{AND}(x_2,x_3), \\ w_4 &= \text{AND}(w_1,w_2), \quad w_5 = \text{AND}(v_1,x_4,x_5) \\ u &= \text{OR}(w_1,w_2,w_3), \quad t = \text{AND}(u,v_2), \quad y = \text{OR}(w_4,t,w_5). \end{aligned}$$

This program has 10 operations and only 24 applications to arguments. To see that it is correct, note that w_4 is true if and only if $x_1 = x_2 = x_3 = 1$, and w_5 is true if and only if $x_4 = x_5 = 1$ and one of x_1,x_2,x_3 is 1. Finally, t is true if and

Figure 4.1
Monotone circuit computing MAJ(x_1,x_2,x_3,x_4,x_5).

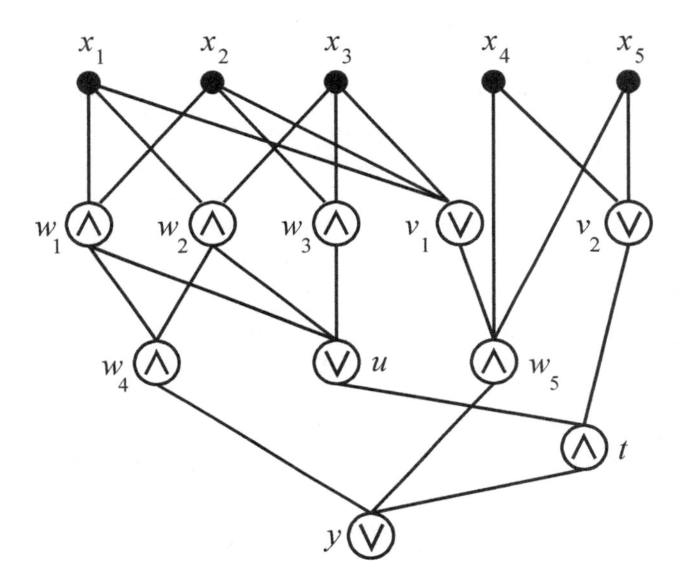

only if two of x_1,x_2,x_3 and one of x_4,x_5 are true, which handles the remaining six true cases.

Now clearly MAJ generalizes into a Boolean function MAJ of strings x of any length n, such that MAJ(x) returns 1 if more than $n/2$ of the bits of x are 1. We ask the important question:

Is MAJ feasible?

The operational question about MAJ(x_1,x_2,x_3,x_4,x_5) is, do the above programs *scale* when "5" is replaced by "n"? Technically, *scalable* is the same idea as *feasible* but with the idea mentioned in chapter 2 that a polynomial bound is the same as having a linear bound each time the size of the input doubles.

The first idea, when generalized from "5" to "n," says to take the AND of every r-sized subset of $[n]$, where $r = \lfloor n/2 \rfloor + 1$, and feed that to an OR.

However, there are $\binom{n}{r}$ such subsets, which is exponential when $r \sim n/2$. So the first idea definitely does not scale.

The trouble with the second, shorter program is its being rather *ad hoc* for $n = 5$. The question of whether there are programs like it with only AND and OR gates that scale for all n is a famous historical problem in complexity theory. The answer is known to be *yes*, but no convenient recipe for constructing the programs for each n is known, and their size $O(n^{5.3})$ is comparatively high.

However, if we think in terms of numbers, we can build circuits that easily scale. Take $k = \lceil \log_2(n+1) \rceil$. We can hold the total count of 1's in an x of length n in a k-bit *register*. So let us mentally draw x going down rather than across and draw to its right an $n \times k$ grid, whose rows will represent successive values of this register. It is nicest to put the least significant bit leftmost, i.e., closest to x, but this is not critical. The row to the right of x_1 has value 0 (that is, 0^k as a Boolean string) if $x_1 = 0$ and value 1 (that is, 10^{k-1}) if $x_1 = 1$. As we scan x downward, if $x_i = 0$, then the row of k bits has the same value as the previous one, but if $x_i = 1$, then the register is incremented by 1. We might not regard the increment operation as *basic*—instead, we might use extra "helper bits" and gates to compute binary addition with carries. But we can certainly tell that we will get a Boolean circuit of size $O(kn) = O(n \log n)$, which is certainly feasible. At the end, we need only compare the final value v with $n/2$, and this is also feasible.

This idea broadens to any computation that we can imagine doing with paper and pencil and some kind of grid, such as multiplication, long division, and other arithmetic. To avoid the scratchwork and carries impinging on our basic grid, we can insist that they occupy h-many "helper rows" below the row with x_n, stretching wires down to those rows as needed. The final idea that helps in progressing to quantum computation is instead of saying we have $k(n+h)$ bits, say that we have $n+h$ bits that "evolve" going left to right. This also fits the classical picture articulated by Turing. The cells directly below x_n in the column with the input can be regarded as a segment of "tape" for scratchwork. The change to this tape at each step of a **Turing Machine** computation on x, including changes to the scratchwork, can be recorded in an adjacent new column. If the machine computation lasts t steps, then with $s = n+h$ as the measure of "space," we have an $s \times t$ grid for the whole computation.

To finish touring machine and circuit models, we can next imagine that every cell in this grid depends on its neighbors in the preceding column by a fixed finite combination of basic Boolean operations. This gives us a circuit of size

$O(st) = O(t^2)$ because we do at most one bit of scratchwork at each step. If the machine time $t(n)$ is polynomial, then so is $t(n)^2$.

This relation also goes in the opposite direction if you think of using a machine to verify the computation by the circuit—provided the circuits are *uniform* in a technical sense that captures the conceptual sense we applied above to Boolean function families.

Hence, the criterion of *feasible* is broad and is the same for any of the classical models of computation by machines, programs, or (uniform) circuits. There is a huge literature on which functions are feasible and which are not. One can encode *anything* via Boolean strings, including circuits themselves. The problem of whether a Boolean circuit can ever output 1—even when it allows applying NOT only to the original arguments x_i and then has just one level of ternary OR gates feeding into a big AND—is not known to be feasible. That is, no feasible function is known to give the correct answer for every encoding X of such a program: this is called the **satisfiability problem** and has a property called **NP-hardness**. We define this with regard to an equivalent problem about solving equations in chapter 16. For now we are happy with not only defining *classical feasible computation* in detail but also showing that equivalent criteria are reached from different models. Now we are ready for the quantum challenge to this standard.

4.3 Quantum Representation of Boolean Arguments

Let $N = 2^n$. Every coordinate in N-dimensional Hilbert space corresponds to a binary string of length n. The **standard encoding scheme** assigns to each index $j \in [0, \ldots, n-1]$ the n-bit binary string that denotes j in binary notation, with leading 0's if necessary. This produces the standard lexicographic ordering on strings. For instance, with $n = 2$ and $N = 4$, we show the indexing applied to a permutation matrix:

	00	01	10	11
00	1	0	0	0
01	0	1	0	0
10	0	0	0	1
11	0	0	1	0

.

The mapping is $f(00) = 00, f(01) = 01, f(10) = 11, f(11) = 10$, and in general $f(x_1,x_2) = (x_1,x_1 \oplus x_2)$. Thus, the operator writes the XOR into the second bit while leaving the first the same. One can also say that it negates the second bit if-and-only-if the first bit is 1. This negation itself is represented on one bit by a matrix we have seen before—now with the indexing scheme, it is:

$$\boldsymbol{X} = \begin{array}{c|cc|} & 0 & 1 \\ \hline 0 & 0 & 1 \\ 1 & 1 & 0 \\ \hline \end{array}.$$

Thus, the negation is controlled by the first bit, which explains the name "Controlled-NOT" (**CNOT**) for the whole 4×4 operation.

To get a general Boolean function $y = f(x_1,\dots,x_n)$, we need $n+1$ Boolean coordinates, which entails $2N = 2^{n+1}$ matrix coordinates. What we really compute is the function

$$F(x_1,\dots,x_n,z) = (x_1,\dots,x_n,z \oplus f(x_1,\dots,x_n)).$$

Formally, F is a Boolean function with outputs in $\{0,1\}^{n+1}$ rather than just $\{0,1\}$. Its first virtue, which is necessary to the underlying quantum physics, is that it is invertible—in fact, F is its own inverse:

$$\begin{aligned} F(F(x_1,\dots,x_n,z)) &= F(x_1,\dots,x_n,z \oplus y) \\ &= (x_1,\dots,x_n,(z \oplus y) \oplus y) = (x_1,\dots,x_n,z). \end{aligned}$$

Its second virtue is having a $2N \times 2N$ permutation matrix $\boldsymbol{P}_f$ that is easy to describe: the lone 1 in each row $x_1x_2\cdots x_nz$ is in column $x_1x_2\cdots x_nb$, where $b = z \oplus f(x_1,\dots,x_n)$.

If f is a Boolean function with m outputs $(y_1,\dots,y_m)$ rather than a single bit, then we have the same idea with

$$F(x_1,\dots,x_n,z_1,\dots,z_m) = (x_1,\dots,x_n,z_1 \oplus y_1,\dots,z_m \oplus y_m)$$

instead. The matrix $\boldsymbol{P}_f$ is still a permutation matrix, although of even larger dimensions $2^{n+m} \times 2^{n+m}$. Often left unsaid is what happens if we need h-many "helper bits" to compute the original f. The simple answer is that we can treat them *all* as extra outputs of the function, allocating extra z_j variables as dummy inputs so that the $\oplus$ trick preserves invertibility. Because h is generally polynomial in n, this does not upset feasibility.

In this scheme, adopting the "rows-and-columns" picture we gave for classical computation above, everything is laid out in $n' = n + m + h$ rows, with the input x laid out in the first column. Each row is said to represent a **qubit**, which is short for *quantum bit*. In order to distinguish the *row* from the idea of a qubit as a physically observable *object*, we often prefer to say **qubit line** for the row itself in the circuit. The h-many helper rows even have their own fancy name as *ancilla* qubits, using the Latin word for "chambermaid" or, more simply, helper.

Writing out a big $2^{n'} \times 2^{n'}$ matrix, just for a permutation, is of course not feasible. This is a chief reason we prefer to think of operators $\boldsymbol{P}_f$ as pieces of code. The qubit lines are really coordinates of binary strings that represent indices to these programs. These strings have size n', and their own indices $1, \dots, n'$ are what we call **quantum coordinates**, when trying to be more careful than saying "qubits." As long as we confine ourselves to linear algebra operations that are efficiently expressible via these n' quantum indices, we can hope to keep things feasible. The rest of the game with quantum computation is, which operations are feasible?

4.4 Quantum Feasibility

A quantum algorithm applies a series of unitary matrices to its start vector. Can we apply any unitary matrix we wish? The answer is no, of course not. If the quantum algorithms are to be efficient, then there must be a restriction on the matrices allowed.

If we look at the matrices $\boldsymbol{P}_f$ in section 4.3, we see several issues. First, the design of $\boldsymbol{P}_f$ seems to take no heed of the complexity of the Boolean function f but merely creates a permutation out of its exponential-sized truth table. Because infeasible (families of) Boolean functions exist, there is no way this alone could *scale*. Second, even for simple functions like $\mathsf{AND}(x_1, x_2, \dots, x_n)$, the matrix still has to be huge—even larger than 2^n on the side. How do we distinguish "basic feasible operations"? Third, what do we use for variables? If we have a 2^n-sized vector, do we need exponentially many variables?

The answer is to note that if we keep the number k of arguments for any operation to a constant, then 2^k stays constant. We can therefore use $2^k \times 2^k$ matrices that apply to just a few arguments. But what *are* the arguments? They are not the same as the Hilbert space coordinates $0, \dots, N-1$, which would involve us in exponentially many. The quantum coordinates start off being labeled $x_1, x_2, \dots, x_n$ as for Boolean input strings and extend to places for outputs and for *ancillae*, which is the plural of *ancilla*.

With these differences understood, the notion of feasible for unitary matrices is the natural extension of the one for Boolean circuits. Any unitary matrix $\boldsymbol{B}$ of dimension 2^k where k is constant—indeed, we will have $k \leq 3$—is feasible. Such a matrix is allowed to operate on any subset of k quantum coordinates, provided it leaves the other $n' - k$ coordinates alone. A tensor product of $\boldsymbol{B}$ with identity matrices on the other quantum coordinates is a **basic matrix**. We could require that the entires of $\boldsymbol{B}$ be simple in some way, but it will suffice to take $\boldsymbol{B}$ from a small fixed finite family of gates.

Now suppose that $\boldsymbol{U}$ is any unitary matrix of dimension N. Then we will say that it is *feasible* provided there is a way to construct it easily out of basic matrices. Technically, we mean that $\boldsymbol{U}$ belongs to an infinite family $[\boldsymbol{U_n}]$ parameterized by n, with each $\boldsymbol{U_n}$ constructible from $n^{O(1)}$ basic matrices in a uniform manner. This stipulation is asymptotic, but the intent is concrete. To show concreteness, one can express $\boldsymbol{U}$ via a **quantum circuit** of basic gate matrices.

We will stay informal with quantum circuits as we did for Boolean circuits while formalizing quantum computations in terms of matrices. Rather than grids with squares as we have described for Boolean circuits, quantum circuits use lines that go across like staves of music and place gates on the lines like musical notes and chords. The first n lines correspond to the inputs $x_1, \ldots, x_n$, while all other qubit lines are conventionally initialized to 0. The only "crossing wires" are parts of multi-ary gates, either running invisibly inside boxes or shown explicitly for some gates like the ***CNOT*** operation above.

Here is a circuit composed of one Hadamard gate on qubit line 1, followed by a ***CNOT*** with its *control* on line 1 and its *target* on line 2:

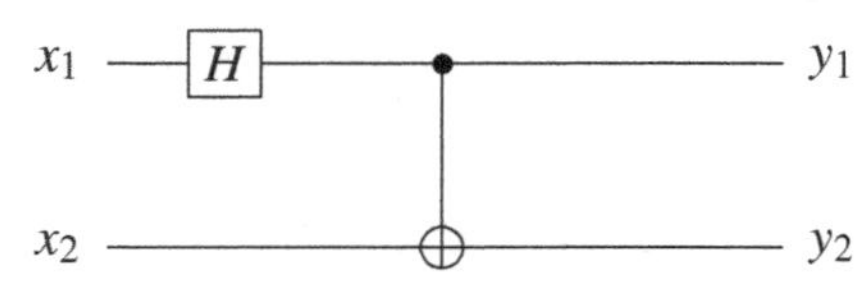

Underneath the Hadamard gate is an invisible identity gate, expressing that in the first time step, the second qubit does not change. We could draw this into the circuit if we wish:

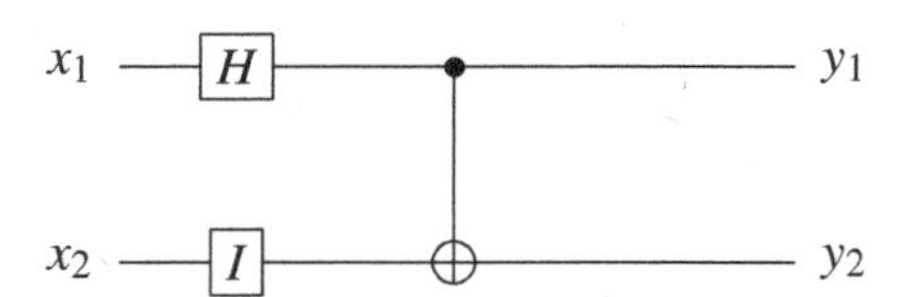

Whenever two gates can be placed vertically this way, a tensor product is involved. Thus, the matrix form of the computation is the composition $\boldsymbol{V_2} \cdot \boldsymbol{V_1}$

of the 4 × 4 matrices

$$\mathbf{V_2} = \mathbf{CNOT} = \begin{bmatrix} 1 & 0 & 0 & 0 \\ 0 & 1 & 0 & 0 \\ 0 & 0 & 0 & 1 \\ 0 & 0 & 1 & 0 \end{bmatrix}, \quad \mathbf{V_1} = \mathbf{H} \otimes \mathbf{I} = \frac{1}{\sqrt{2}} \begin{bmatrix} 1 & 0 & 1 & 0 \\ 0 & 1 & 0 & 1 \\ 1 & 0 & -1 & 0 \\ 0 & 1 & 0 & -1 \end{bmatrix}.$$

To show their product $\mathbf{U} = \mathbf{V_2 V_1}$ acting on the input vector $x = [1,0,0,0]^T$, we obtain

$$\frac{1}{\sqrt{2}} \begin{bmatrix} 1 & 0 & 1 & 0 \\ 0 & 1 & 0 & 1 \\ 0 & 1 & 0 & -1 \\ 1 & 0 & -1 & 0 \end{bmatrix} \begin{pmatrix} 1 \\ 0 \\ 0 \\ 0 \end{pmatrix} = \frac{1}{\sqrt{2}} \begin{pmatrix} 1 \\ 0 \\ 0 \\ 1 \end{pmatrix}. \tag{4.1}$$

Expressed in quantum coordinates, the input vector denotes the input string 00, that is, $x_1 = 0$ and $x_2 = 0$. The output vector is not a basis vector. Rather, it is an equal-weighted sum of the basis vector for 00 and the basis vector for 11. This means we do not have a unique output string y in quantum coordinates, although we have a simple output vector v in the $N = 2^2 = 4$ Hilbert-space coordinates. Much of the power of quantum algorithms will come from such outputs v, from which we need further interaction in the form of *measurements*, even repeatedly, to arrive at a final Boolean output y. Thus, we hold off saying what it means for a Boolean function to be quantum feasibly computable, but we have enough to define formally what it means for a quantum *computation* to be feasible:

DEFINITION 4.2 A quantum computation $\mathbf{C}$ on s qubits is **feasible** provided

$$\mathbf{C} = \mathbf{U}_t \mathbf{U}_{t-1} \cdots \mathbf{U}_1,$$

where each $\mathbf{U}_i$ is a feasible operation, and s and t are bounded by a polynomial in the designated number n of input qubits.

Which quantum gates $\mathbf{B}$ are basic, and which possibly other operations $\mathbf{V}$ on the whole space are feasible? We will not try to give a comprehensive answer to this question, but in the next chapter, we give some more gates that everyone agrees are basic and some operations that almost everyone agrees are feasible. They are building blocks in the same way that the Boolean operations NOT, AND, OR, and XOR are. It is believed—certainly hoped—that such matrices will one day be constructible. Moreover, quantum computers may prove to be makable out of replicable parts the way classical computers—such as your laptop—are built today.

4.5 Problems

4.1. Show that XOR cannot be written using only the composition of AND and OR functions.

4.2. Suppose that $f(x)$ and $g(x)$ are Boolean functions on n inputs. Let

$$h(x) = f(x) \oplus g(x).$$

Prove that h is always zero if and only if f and g are the same function.

4.3. For a Boolean string $x = x_1, \ldots, x_n$ define

$$(-1)^x$$

to be

$$(-1)^{(x_1 + \cdots + x_n)}.$$

Show that $(-1)^x$ is equal to 1 if and only if $\mathsf{XOR}(x) = 0$.

4.4. Let

$$x = x_1, \ldots, x_n,$$

and

$$y = y_1, \ldots, y_n,$$

be Boolean strings. Prove that

$$(-1)^{x \oplus y} = (-1)^x \times (-1)^y.$$

4.5. Does $(-1)^{x \bullet y}$ always equal $(-1)^{x \oplus y}$?

4.6. Show that there are an uncountable number of unitary matrices of dimension N. Does this help explain why we cannot allow any unitary matrix?

4.7. Show that $x_1 \oplus \cdots \oplus x_n$ can be formed from $O(n)$ binary Boolean operations.

4.8. What effect does the following 4×4 matrix have on the quantum coordinates—that is, on the four basis vectors for the strings 00, 01, 10, and 11?

$$\begin{bmatrix} 1 & 0 & 0 & 0 \\ 0 & 0 & 1 & 0 \\ 0 & 1 & 0 & 0 \\ 0 & 0 & 0 & 1 \end{bmatrix}$$

We include it as a basic gate.

4.9. Now show how to use the *swap gate* of problem 4.8 to obtain the following action on the standard basis vectors: the string $x_1x_2 \cdots x_n$ becomes $x_2 \cdots x_nx_1$. Call this the **cycle $\boldsymbol{K_n}$**. How many swap gates did you need?

4.10. Consider a quantum circuit with three qubit lines. We can draw a ***CNOT*** gate whose *control* is separated from its *target*—indeed, we can also place it "upside down" as:

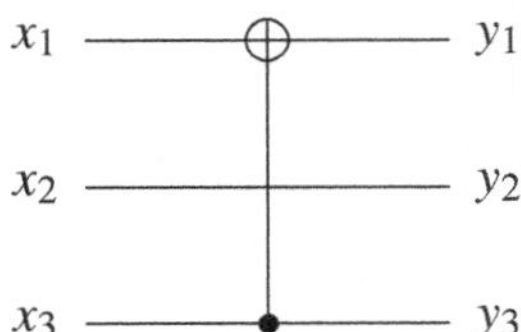

We have omitted showing an identity gate on the second qubit line, which would look ugly when crossed by the ***CNOT*** gate's vertical wire anyway. Write out the 8×8 matrix of the operation represented by this (piece of a) quantum circuit.

4.11. Show that the matrix in problem 4.10 *cannot* be written as a tensor product of two smaller matrices, in particular not of some permutation of ***CNOT*** and the 2×2 identity matrix. This seems to violate our definition of a basic matrix $\boldsymbol{V_i}$ in a feasible quantum computation, but see the next problem.

4.12. Show nevertheless that the circuit in problem 4.10 can be simulated by three quantum time-steps, each a tensor product of the 2×2 identity and a basic 4×4 matrix. Argue generally that k-qubit gates can be oriented in any way desired on any k qubit lines without upsetting definition 4.2. *Hint:* Use problem 4.8.

4.13. Consider the general construction in section 4.3 of a unitary permutation matrix $\boldsymbol{P_f}$ from a Boolean function f. For what function f does ***CNOT*** equal $\boldsymbol{P_f}$?

4.14. Determine the 8×8 matrix of $\boldsymbol{P}_f$ where f is the binary AND function. (You do not have to write every 0.) This gate, which we write as ***TOF***, is named for Tommaso Toffoli.

4.6 Summary and Notes

This chapter has presented the basic ingredients of Boolean complexity and quantum complexity side by side for comparison. In both cases, there is a common notion of *feasible* associated with complexity cost measures being bounded by some polynomial in the size of the data. We have presented the Boolean circuit model in both of its equivalent formulation via circuits and straight-line programs, and while we support viewing quantum computations as circuits, we defined them as programs giving compositions of basic matrix operations.

There are many books on Boolean complexity—check one out if you need more background here. General textbooks on computation theory also include concepts such as machine models, decision problems, and (un)computability. Among them we suggest the texts by Sipser (2012) and Homer and Selman (2011). The second author has co-written three book chapters on the basics of complexity theory (Allender et al., 2009a,b,c). The first of these chapters includes a diagram of the simulation of Turing machines by Boolean circuits (with st size overhead) in the form of Savage (1972); an $O(t \log s)$-size simulation was proved by Pippenger and Fischer (1979), but these circuits do not have the same degree of spatial locality. The theorem about monotone programs for majority was proved by Valiant (1984). That quantum operations can simulate Turing machines was first observed by Benioff (1982).

The ***CNOT*** gate and some other quantum gates go all the way back to Feynman (1982, 1985) and Deutsch (1985, 1989), while Yao (1993) systematized quantum circuit theory, and Barenco et al. (1995) gave an influential roundup of basic gates. Universality results about small gate sets followed (Barenco et al., 1995, DiVincenzo, 1995, Lloyd, 1995). The Toffoli gate comes from Toffoli (1980) and Fredkin and Toffoli (1982).

5 Special Matrices

Given our view that quantum algorithms are simply the result of applying a unitary transformation to a unit vector, it should come as no surprise that we need to study unitary matrices. Happily, there are just a few families of such matrices that are used in most quantum algorithms. We will present those in this chapter.

Two of the families correspond to transforms that are well studied through mathematics and computer science theory and have many applications in many areas besides quantum algorithms. When is a transformation a *transform*? The latter term connotes that the output is a new way of interpreting the input. Because all quantum transformations are invertible, this is in a sense always true, but the intuition is highest for the families presented here.

5.1 Hadamard Matrices

The first family of unitary transforms are the famous Hadamard matrices. Note that because we mainly stay with the standard basis of $\mathbf{e}_k$ vectors, we will identify transforms with their matrices, and this should cause no confusion. Here we lock in our convention that N is always 2^n for some n.

DEFINITION 5.1 The Hadamard matrix $\mathbf{H}_N$ of order N is recursively defined by $\mathbf{H}_2 = \mathbf{H}$ and for $N \geq 4$:

$$\mathbf{H}_N = \mathbf{H}_{N/2} \otimes \mathbf{H} = \frac{1}{\sqrt{2}} \begin{bmatrix} \mathbf{H}_{N/2} & \mathbf{H}_{N/2} \\ \mathbf{H}_{N/2} & -\mathbf{H}_{N/2} \end{bmatrix}$$

We could also use $\mathbf{H}_1 = [1]$ as the basis. If we wish to use n not N as a marker, then we write $\mathbf{H}^{\otimes n}$ using a superscript instead of a subscript.

This recursive definition implies many important facts about this matrix. For example, it easily implies that, in general, $\mathbf{H}_N$ is equal to $\frac{1}{\sqrt{N}}\mathbf{A}$, where $\mathbf{A}$ is a matrix of ± 1 only. However, it is often much more useful to have the following direct definition of the entries of $\mathbf{H}_N$.

LEMMA 5.2 For any row r and column c,

$$\mathbf{H}_N[r,c] = (-1)^{r \bullet c},$$

recalling that $r \bullet c$ is the inner product of r and c treated as Boolean strings. □

Thus, for any vector $\boldsymbol{a}$, the vector $\boldsymbol{b} = \boldsymbol{H}_N\boldsymbol{a}$ is defined by

$$\boldsymbol{b}(x) = \frac{1}{\sqrt{N}} \sum_{t=0}^{N-1} (-1)^{x \bullet t} \boldsymbol{a}(t).$$

This is the way that we will view the transform in the analysis of most algorithms. Note the convenience of using Boolean strings as index arguments.

In a quantum circuit with n qubit lines, $\boldsymbol{H}_N$ is shown as a column of n-many single-qubit Hadamard gates. This picture frees one from having to think of tensor products in the design of a circuit but does not further our analysis.

5.2 Fourier Matrices

The next important family consists of the quantum Fourier matrices. Let ω stand for $e^{2\pi i/N}$, which is often called "the" principal N-th root of unity.

DEFINITION 5.3 The Fourier matrix $\boldsymbol{F}_N$ of order N is:

$$\frac{1}{\sqrt{N}} \begin{bmatrix} 1 & 1 & 1 & 1 & \cdots & 1 \\ 1 & \omega & \omega^2 & \omega^3 & \cdots & \omega^{N-1} \\ 1 & \omega^2 & \omega^4 & \omega^6 & \cdots & \omega^{N-2} \\ 1 & \omega^3 & \omega^6 & \omega^9 & \cdots & \omega^{N-3} \\ \vdots & & & \vdots & \ddots & \vdots \\ 1 & \omega^{N-1} & \omega^{N-2} & \omega^{N-3} & \cdots & \omega \end{bmatrix}$$

That is, $\boldsymbol{F}_N[i,j] = \omega^{ij \bmod N}$ divided by $\sqrt{N}$.

It is well known that $\boldsymbol{F}_N$ is a unitary matrix over the complex Hilbert space. This and further facts about $\boldsymbol{F}_N$ are set as exercises at the end of this chapter, including a running theme about its feasibility via various decompositions. For any vector $\boldsymbol{a}$, the vector $\boldsymbol{b} = \boldsymbol{F}_N\boldsymbol{a}$ is defined in our index notation by:

$$\boldsymbol{b}(x) = \frac{1}{\sqrt{N}} \sum_{t=0}^{N-1} \omega^{xt} \boldsymbol{a}(t).$$

This is the way that we will view the transform in our algorithmic analysis. That this is tantalizingly close to the equation for the Hadamard transform was significant to Peter Shor in his step from Daniel Simon's algorithm (chapter 10) to his own (chapter 11). The differences are having ω in place of -1 and multiplication xt in place of the Boolean inner product $x \bullet t$.

5.3 Reversible Computation and Permutation Matrices

Every $N \times N$ permutation matrix is unitary. However, in terms of n with $N = 2^n$, there are doubly exponentially many permutation matrices. Hence, not all of them can be feasible—indeed, most of them are concretely infeasible. Which permutation matrices are feasible?

We can give a partial answer. Recall the definition of the permutation matrix $\boldsymbol{P}_f$ from the invertible extension F of a Boolean function f in section 4.3.

THEOREM 5.4 All classically feasible Boolean functions f have feasible quantum computations in the form of $\boldsymbol{P}_f$.

The proof of this theorem stays entirely classical—that is, the quantum circuits are the same as Boolean circuits that are **reversible**, which in turn efficiently embed any given Boolean circuit computing f. We need only one new gate, which was already mentioned in problem 4.14.

DEFINITION 5.5 The **Toffoli gate** is the ternary Boolean function

$$\boldsymbol{TOF}(x_1, x_2, x_3) = (x_1, x_2, x_3 \oplus (x_1 \wedge x_2)).$$

The Toffoli gate induces the permutation in 8-dimensional Hilbert space that swaps the last two entries, which correspond to the strings 110 and 111, and leaves the rest the same. This extends the idea of $\boldsymbol{CNOT}$ with x_1, x_2 as "controls" and x_3 as the "target." That this simple swap is universal for Boolean computation is conveyed by the following two facts for Boolean bit arguments a, b:

- $\text{NOT}(a) = \boldsymbol{TOF}(1, 1, a)$;
- $\text{AND}(a, b) = \boldsymbol{TOF}(a, b, 0)$.

Proof of Theorem 5.4. Because AND and NOT is a universal set of logic gates, we may start with a Boolean circuit C computing $f(x_1, \ldots, x_n)$ using r-many NOT and s-many binary AND gates. The NOT gates we can leave alone because we already have the corresponding 2×2 matrix $\boldsymbol{X}$ as a basic quantum operation. Hence, we need only handle the s-many AND gates. We can simulate them by s-many Toffoli gates each with an ancilla line set to 0 for input, but this is superseded by the issue of possibly needing multiple copies of the result c of an AND gate on lines a, b—that is, one for each *wire* out of the gate.

This is where the Toffoli gate shines. For each output wire w, we allocate a fresh ancilla z and put a Toffoli gate with target on line z and controls on a and b. This automatically computes $z \oplus (a \wedge b)$, which with z initialized to 0 is what we want. Multiple Toffoli gates with the same controls do not affect each other. Hence, the overhead is bounded by the number of *wires* in C, which is polynomial, and the only ancilla lines we need already obey the convention of being initialized to 0. □

There are versions of theorem 5.4, some applying to Turing machines and other starting models of computation, that have much less overhead, but "polynomial" is good enough for our present discussion of feasibility. Thus, a permutation matrix—which is a deterministic quantum operation—is feasible if it is induced by a classical feasible function on the quantum coordinates.

5.4 Feasible Diagonal Matrices

Any diagonal matrix whose entries have absolute value 1 is unitary. Hence, it can be a quantum operation. The question is, which of these operations are *feasible*?

Of course if the size of the matrix is a small fixed number, we can call it basic and hence feasible. What happens when the matrices are $N \times N$, however? Even if we limit to entries 1 and -1, we have one such matrix $\boldsymbol{U}_S$ for every subset S of $[N]$, that is, $S \subseteq \{0,1\}^n$:

$$\boldsymbol{U}_S[x,x] = \begin{cases} -1 & \text{if } x \in S; \\ 1 & \text{otherwise.} \end{cases}$$

Because there are doubly exponentially many S, there are doubly exponentially many $\boldsymbol{U}_S$, so most of them are not feasible. But can we tell which *are* feasible? Again we give a partial answer. When S is the set of arguments that make a Boolean function f true, we write $\boldsymbol{U}_f$ in place of $\boldsymbol{U}_S$. The matrix $\boldsymbol{U}_f$ is called the **Grover oracle** for f.

THEOREM 5.6 If f is a feasible Boolean function, then its Grover oracle $\boldsymbol{U}_f$ is feasible.

We defer the proof until section 6.5 in the next chapter. As with theorem 5.4, the question of whether any *other* families of functions f make $\boldsymbol{U}_f$ meet our quantum definition of feasible is a deep one whose answer is long unknown and is related to issues in chapter 16. We can be satisfied for now that we have a rich vocabulary of feasible operations, and the next chapter will give some tricks for combining them. Here we give one more family of operations.

5.5 Reflections

Given any unit vector $\boldsymbol{a}$, we can create the unitary operator $\mathbf{Ref}_{\boldsymbol{a}}$, which *reflects* any other unit vector $\boldsymbol{b}$ around $\boldsymbol{a}$. Geometrically, this is done by dropping a line from the tip of $\boldsymbol{b}$ that hits the body of $\boldsymbol{a}$ in a right angle and continuing the line the same distance further to a point $\boldsymbol{b}'$. Then $\boldsymbol{b}'$ likewise lies on the unit sphere of the Hilbert space. The operation mapping $\boldsymbol{b}$ to $\boldsymbol{b}'$ preserves the unit sphere and is its own inverse, so it is unitary.

In geometrical terms, the point on the body of $\boldsymbol{a}$ is the **projection** of $\boldsymbol{b}$ onto $\boldsymbol{a}$ and is given by $\boldsymbol{a}' = \boldsymbol{a}\langle \boldsymbol{a}, \boldsymbol{b} \rangle$. Thus,

$$\boldsymbol{b}' = \boldsymbol{b} - 2(\boldsymbol{b} - \boldsymbol{a}\langle \boldsymbol{a}, \boldsymbol{b} \rangle) = (2\mathbf{P}_{\boldsymbol{a}} - \mathbf{I})\boldsymbol{b},$$

where $\mathbf{P}_{\boldsymbol{a}}$ is the operator doing the projection: for all $\boldsymbol{b}$,

$$\mathbf{P}_{\boldsymbol{a}}\boldsymbol{b} = \boldsymbol{a}\langle \boldsymbol{a}, \boldsymbol{b} \rangle.$$

For example, let $\boldsymbol{a}$ be the unit vector with entries $\frac{1}{\sqrt{N}}$, which we call $\boldsymbol{j}$. Then the projector is the matrix whose entries are all $\frac{1}{N}$, which we call $\mathbf{J}$ in our matrix font. Finally, the reflection operator is

$$\mathbf{V} = 2\mathbf{J} - \mathbf{I} = \begin{bmatrix} \frac{2}{N} - 1 & \frac{2}{N} & \cdots & \\ \frac{2}{N} & \frac{2}{N} - 1 & \cdots & \\ \frac{2}{N} & \frac{2}{N} & \cdots & \ddots \end{bmatrix}.$$

We claim this matrix is feasible. Of course this leads to the question: which reflection operations are feasible?

An important case of reflection is when $\boldsymbol{a}$ is the *characteristic vector* of a nonempty set S, that is:

$$\boldsymbol{a}(x) = \begin{cases} \frac{1}{\sqrt{|S|}} & \text{if } x \in S; \\ 0 & \text{otherwise.} \end{cases}$$

Suppose we apply $\mathbf{Ref}_{\boldsymbol{a}}$ to vectors $\boldsymbol{b}$ with the foreknowledge that all entries $e = \boldsymbol{b}(x)$ for $x \in S$ *are equal*. Let $k = |S|$. Then we have $\langle \boldsymbol{a}, \boldsymbol{b} \rangle = ke/\sqrt{k} = e\sqrt{k}$, and taking the projection $\boldsymbol{a}' = \boldsymbol{P}_a\boldsymbol{b}$, we have

$$\boldsymbol{a}'(x) = \begin{cases} e & \text{if } x \in S; \\ 0 & \text{otherwise.} \end{cases}$$

The reflection $\boldsymbol{b}' = 2\boldsymbol{a}' - \boldsymbol{b}$ thus satisfies

$$\boldsymbol{b}'(x) = \begin{cases} \boldsymbol{b}(x) & \text{if } x \in S; \\ -\boldsymbol{b}(x) & \text{otherwise,} \end{cases}$$

because in the case $x \in S$, $\boldsymbol{b}'(x) = 2e - \boldsymbol{b}(x) = 2e - e = e = \boldsymbol{b}(x)$. Then the action is the same as multiplying by the diagonal matrix that has -1 for the coordinates that are *not* in S, that is, by the Grover oracle for the complement of S. Because the negation of a feasible Boolean function is feasible, this together with the case of $\boldsymbol{V}$ implies:

THEOREM 5.7 For all feasible Boolean functions f, provided we restrict to the linear subspace of argument vectors whose entries indexed by the "true set" S_f of f are equal, reflection about the characteristic vector of S_f is a feasible quantum operation. □

Happily, the set of such argument vectors forms a linear subspace and always contains the vector $\boldsymbol{j}$, which we will use as a "start" vector. Moreover, reflections by $\boldsymbol{a}$ and $\boldsymbol{b}$, when applied to vectors already in the linear subspace spanned by $\boldsymbol{a}$ and $\boldsymbol{b}$, stay within that subspace. We will use this when presenting Grover's algorithm and search by quantum random walks, but that is getting ahead of our story.

5.6 Problems

5.1. What is $\boldsymbol{H}_4$?

5.2. Prove that $\boldsymbol{H}_N$ is a unitary matrix.

5.3. Prove lemma 5.2.

5.4. Prove that $\boldsymbol{F}_N$ is a unitary matrix. Note that because it is a complex valued matrix, this means that it times its complex transpose is the identity matrix.

5.5. Let $\boldsymbol{D}_N$ be the diagonal matrix formed by normalizing the top half of the second column of $\boldsymbol{F}_{2N}$. Show how to write $\boldsymbol{D}_N$ as a tensor product of the 2×2 twist matrices $\boldsymbol{T}_\alpha$ defined in problem 3.10.

5.6. Recalling also the cycle operation $\boldsymbol{K}_n$ from problem 4.9, show that $\boldsymbol{F}_N$ obeys the following recursive equation in block matrices:

$$\boldsymbol{F}_N = \frac{1}{\sqrt{2}} \begin{bmatrix} \boldsymbol{I}^{\otimes(n-1)} & \boldsymbol{D}_{N/2} \\ \boldsymbol{I}^{\otimes(n-1)} & -\boldsymbol{D}_{N/2} \end{bmatrix} \begin{bmatrix} \boldsymbol{F}_{N/2} & 0 \\ 0 & \boldsymbol{F}_{N/2} \end{bmatrix} \boldsymbol{K}_n^{-1}.$$

This ultimately shows how to decompose $\boldsymbol{F}_N$ into swaps, controlled twists, and the Hadamard gates at the base of this recursion.

5.7. Let $f\colon \{0,1\}^n \to \{0,1\}^n$ be a Boolean function. Show that the following function is always invertible:

$$g(x,y) = (x, y \oplus f(x)).$$

5.8. Prove that the following sum is zero:

$$\sum_{k=0}^{N-1} \omega^k,$$

where ω is $e^{2\pi i/N}$. For what values of ℓ is the following sum equal to zero:

$$\sum_{k=0}^{N-1} \omega^{k\ell}?$$

5.9. Show that the product of two permutation matrices is again a permutation matrix. Also show that a permutation matrix is unitary.

5.10. A *Fredkin gate*, named for Edward Fredkin, swaps 101 and 110 while leaving the other six arguments fixed. Show that, like the Toffoli gate, it is universal for reversible computation.

5.11. Show by direct calculation that the reflection matrix $2\boldsymbol{J} - \boldsymbol{I}$ is unitary.

5.12. Our use of the term *characteristic vector* in section 5.5 may appear to clash with the standard term in linear algebra, which we prefer to call an *eigenvector*. Show, however, that the meanings do harmonize, namely, the characteristic vector is an eigenvector of some relevant quantum operations.

For the following problems, define two vectors to be *dyadically orthogonal* if their entry-wise products cancel in pairs. That is, two vectors $\boldsymbol{a}, \boldsymbol{b}$ in $\mathbb{C}^n$, where n is necessarily even, are dyadically orthogonal if $[n]$ can be partitioned into two-element subsets $\{i,j\}$ such that $\overline{a(i)}b(i) = -\overline{a(j)}b(j)$. Of course two dyadically orthogonal vectors are orthogonal. Here are some pairs of dyadically orthogonal real and complex vectors:

$$\begin{matrix}[1 & 1 & 1 & 1] \\ [1 & 1 & -1 & -1]\end{matrix}, \quad \begin{matrix}[9 & 4 & 6 & -15] \\ [5 & -3 & 2 & 3\,]\end{matrix}, \quad \begin{matrix}[1 & i & 1 & 1] \\ [\,i & 1 & 1 & -1]\end{matrix}.$$

Next, call a matrix *dyadically unitary* if it is unitary and, in addition, every two distinct rows are dyadically orthogonal and likewise every two distinct columns. Every 2×2 unitary matrix is dyadically unitary, so the concept becomes distinctive starting with 4×4 matrices.

5.13. Show that every Hadamard matrix $\boldsymbol{H}_N$ is dyadically unitary.

5.14. Show that the quantum Fourier transform matrices $\boldsymbol{F}_N$ are dyadically unitary.

5.15. Show that the tensor product of a dyadically unitary matrix with *any* unitary matrix is dyadically unitary.

5.16. Show that dyadic unitarity is (alas) not closed under composition, that is, under matrix product. In particular, find a 4×4 dyadically unitary matrix $\boldsymbol{A}$ such that $\boldsymbol{A}^2$, while necessarily unitary, is not dyadically so. (Hint: First do problem 3.19 in chapter 3.)

The following exercises give more understanding of quantum coordinates and embody a research avenue we two authors began toward deeper analysis of the quantum Fourier transform. We generalize the notion of substring for any subset I of $\{1,\dots,n\}$, $I = \{i_1, i_2, \dots, i_r\}$ in order, by putting $x_I = x_{i_1}x_{i_2}\cdots x_{i_r}$, for any $x \in \{0,1\}^n$. For any such I and binary string w of length r, define

$$S_{I,w} = \{x \mid x_I = w\}.$$

Under the standard binary order of complex indices, we regard $S_{I,w}$ also as a subset of $\{0,\dots,N-1\}$ (where $N = 2^n$ as usual) and call $S_{I,w}$ a *cylinder*. Sets of the form $S_I = S_{I,0^r}$, where $r = |I|$, are *principal cylinders*.

5.17. Show that for any r, the first $R = 2^r$ complex indices form a principal cylinder.

5.18. Write out the members of the cylinder for $n = 5$, $I = \{2,3,5\}$, and $w = 101$. How can we recognize the numbers in binary notation?

Now define an $N \times N$ matrix $\boldsymbol{M}$ to be **unitarily decomposable** along a set S of rows, where $R = |S|$ divides N, if the columns of $\boldsymbol{M}$ can be partitioned into N/R-many subsets $T_0, \ldots, T_{N/R-1}$, each of size R, such that for all $j < N/R$, the $R \times R$ sub-matrix $\boldsymbol{M}[S, T_j]$ (which is formed by the rows in S and the columns in T_j) is unitary up to a scalar multiple. The "blocks" T_j need not consist of consecutive columns.

5.19. Show that for any principal cylinder S, the Hadamard matrices $\boldsymbol{H}_N$ are unitarily decomposable along S, where the sub-matrices are also Hadamard matrices up to factors of $\sqrt{2}$. Is this true of any cylinder? Find a set of four rows in $\boldsymbol{H}_8$ that have no unitary decomposition.

5.20. Show that for any principal cylinder S, the quantum Fourier transform $\boldsymbol{F}_N$ is unitarily decomposable along S.

Hint: The proof we know works by induction on $N = 2^n$, dividing into cases according to whether the nth qubit belongs to I. For this we have found it convenient to prove and maintain the stronger inductive hypothesis that every submatrix $\boldsymbol{U}_j = \boldsymbol{M}[S, T_j]$ is dyadically unitary and, moreover, that they are all related by powers of a unitary diagonal matrix $\boldsymbol{D}$ on the left. That is, $\boldsymbol{U}_j = D^{p(j)}\boldsymbol{U}_0$, where however the powers $p(j)$ need not be consecutive integers.

5.7 Summary and Notes

This chapter has presented the most important operations for quantum computation. It has given us a vocabulary of feasible quantum operations. In particular:

1. The special matrices Hadamard and Fourier are all feasible.
2. Permutation matrices allowed by the reversibility theorem are feasible provided the corresponding Boolean function is classically feasible.
3. The Grover oracle of a classically feasible Boolean function is feasible.

A curious piece of history is that the family we call the Hadamard matrices were not discovered by Jacques Hadamard but by Joseph Sylvester. The transform is also named for Joseph Walsh and/or Hans Rademacher. Mathematical history can be complex.

The important results on reversible computation are due to Bennett (1973) and Lecerf (1963), who discovered them independently years before quantum algorithms were even envisioned. Their motivation was to study the limits of reversible computations for their own sake. At one time, it was thought that computations had to destroy information: for example, every assignment operation destroys the previous contains of a memory location and so causes a loss of information. Now we know that any computation can be made reversible, the destruction of information is not required for computation, and, even better, reversibility does not greatly increase computational cost. The Fredkin gate is from Fredkin and Toffoli (1982).

Dyadic unitarity is an original concept whose point is realized in problem 5.20. The theorem there about decompositions of QFT is original work by the second author.

Tricks

There are several tricks of the trade, that is, tricks that are used in quantum algorithms, which are so "simple" they are rarely explained in any detail. We will not do that. We will give you the secrets of all the tricks so you can become a master. Okay, at least you will be able to follow the algorithms that we will soon present.

6.1 Start Vectors

A quantum algorithm needs to start on a simple vector. Just like classical algorithms, we usually restrict algorithms to start in a simple state. It may be okay to assume that all memory locations are set to zero, but it is usually not okay to assume that memory contains the first m primes. We have the same philosophy: start states must be simple.

The simplest start state possible is $\boldsymbol{e}_0 = [1,0,0,\ldots,0]$. This is the one we would like generally to start with, but there are exceptions. Indeed, the first algorithm we will shortly present starts up in $\boldsymbol{e}_1 = [0,1,0,0]$. Because this is also an elementary vector, it is reasonable to allow this as the initial state. An alternative is to show how to move from $\boldsymbol{e}_0$ to $\boldsymbol{e}_1$ in a manner independent of the dimension N.

The idea is that with respect to the indexing scheme, 0 corresponds to the string 0^n and 1 to $0^{n-1}1$, which differ only in the least place. Hence, we can regard the change as **local** with respect to the indexing of strings, which in turn corresponds to the ordering in tensor products. Thus, inverting the last bit, which entails swapping $\boldsymbol{e}_0$ and $\boldsymbol{e}_1$, is accomplished by tensoring $\boldsymbol{I}^{\otimes(n-1)}$ with the matrix $\boldsymbol{X}$ we saw in chapter 3,

$$\boldsymbol{X} = \begin{bmatrix} 0 & 1 \\ 1 & 0 \end{bmatrix}.$$

Although this creates a huge matrix and looks heavy, it is just the linear-algebraic way of applying a NOT gate to the last string index. We can do this on the rth bit from the right, inducing permutations of $[N]$ that move indices up or down by 2^r. Note that we have not transposed *only* $\boldsymbol{e}_0$ and $\boldsymbol{e}_1$; we must be aware of other effects on the Hilbert space.

Interchanging $\boldsymbol{e}_1$ and $\boldsymbol{e}_2$ involves a different operation. In string indices, we need to swap $\ldots 01$ with $\ldots 10$. This is not totally local as it involves two indices, but nearly so. Now we need to tensor the 4×4 **swap** matrix,

$$\mathbf{SWAP} = \begin{bmatrix} 1 & 0 & 0 & 0 \\ 0 & 0 & 1 & 0 \\ 0 & 1 & 0 & 0 \\ 0 & 0 & 0 & 1 \end{bmatrix}$$

after $\mathbf{I}^{\otimes(n-2)}$. This can be regarded as a benefit of the binary function $f(a,b) = (b,a)$ being invertible.

Another interesting start vector $\boldsymbol{j}$ is the sum of all the $\boldsymbol{e}_k$, which must be divided by $\sqrt{N}$ to keep it a unit vector. Aside from this normalizing factor, it has a 1 in each entry. We can obtain this from $\boldsymbol{e}_0$ by noting that the Hadamard matrix $\mathbf{H}_N = \mathbf{H}^{\otimes n}$ has 1's in its entire left column, and moreover it comes with the same $\sqrt{N} = 2^{n/2}$ factor. So we have

$$\boldsymbol{j}_N = \mathbf{H}_N \boldsymbol{e}_0.$$

Again by the tensor product feature, this operation is local to each individual string index. In terms of strings, it creates a weighted sum over all of $\{0,1\}^n$. If we apply this to any other $\boldsymbol{e}_k$, then we get a vector with some -1 entries in place of $+1$ but giving the same squared amplitudes. This is because other columns in $\mathbf{H}_N$ have negative entries.

Finally, we may wish to extend our start vectors to initialize helper bits. Generally, this means extending the underlying binary string with some number m of 0s. In that case, because we already regard $\boldsymbol{e}_0$ as our generic start vector, we need do nothing. Algebraically what we are doing is working in the product Hilbert space $\mathbb{H}_N \otimes \mathbb{H}_M$, with $M = 2^m$, because $\boldsymbol{e}_0$ in the product space is just the tensor product of the first basis vectors of the two spaces. If we want to change any state $\boldsymbol{a}$ to $\boldsymbol{a} \otimes \boldsymbol{e}_0$, then we may suppose the extra helper bits were there all along.

6.2 Controlling and Copying Base States

Can we change any state $\boldsymbol{a}$ to $\boldsymbol{a} \otimes \boldsymbol{a}$? Algebraically, the latter means the state $\boldsymbol{b}$ such that indexing by strings $x, y \in \{0,1\}^n$, we have

$$\boldsymbol{b}(xy) = \boldsymbol{a}(x)\boldsymbol{a}(y).$$

The famous **no-cloning theorem** says that there is **no** $2^{2n} \times 2^{2n}$ unitary operation $\mathbf{U}$ such that for all $\mathbf{a}$,

$$\mathbf{U}(\mathbf{a} \otimes \mathbf{e}_0) = \mathbf{a} \otimes \mathbf{a}.$$

However, a limited kind of copying is possible that can replicate computations and help to *amplify* the success probability of algorithms after taking measurements.

THEOREM 6.1 For any $n \geq 1$, we can efficiently build a $2^{2n} \times 2^{2n}$ unitary operation $\mathbf{C_n}$ that converts any vector $\mathbf{a}'$ into $\mathbf{b}$ such that for all $x, y \in \{0,1\}^n$,

$$\mathbf{b}(xy) = \mathbf{a}'(x(x \oplus y)).$$

In particular, if $\mathbf{a}' = \mathbf{a} \otimes \mathbf{e}_{0^n}$, then we get for all x,

$$\mathbf{a}(x) = \mathbf{a}'(x0^n) = \mathbf{b}(xx),$$

so that measuring $\mathbf{b}$ yields xx with the same probability that measuring $\mathbf{a}$ yields x.

Proof. First consider $n = 1$. The operator must make $\mathbf{b}(00) = \mathbf{a}(00)$, $\mathbf{b}(01) = \mathbf{a}(01)$, $\mathbf{b}(10) = \mathbf{a}(11)$, and $\mathbf{b}(11) = \mathbf{a}(10)$. This is done by the 4×4 permutation matrix

$$\mathbf{CNOT} = \begin{bmatrix} 1 & 0 & 0 & 0 \\ 0 & 1 & 0 & 0 \\ 0 & 0 & 0 & 1 \\ 0 & 0 & 1 & 0 \end{bmatrix}.$$

As we saw in section 4.3, the name $\mathbf{CNOT}$ stands for "Controlled-Not" because the second qubit is negated if the first qubit has a 1 value and is left unchanged otherwise.

For $n = 2$, the indices are length-4 strings $y_1 y_2 z_1 z_2$, which are permuted into $y_1 y_2 (y_1 \oplus z_1)(y_2 \oplus z_2)$. This is a composition of two $\mathbf{CNOT}$ operations, one on the first and third indices (preserving the others), which we denote by $\mathbf{C_{1,3}}$, and the other on the second and fourth, written as $\mathbf{C_{2,4}}$. For general n, the final operator is the composition $\mathbf{C_n} = \mathbf{C_{1,n+1}} \mathbf{C_{2,n+2}} \cdots \mathbf{C_{n,2n}}$. □

Magically, what this does is clone every basis state at once. If $\mathbf{a} = \mathbf{e}_x$, then $\mathbf{b}$ is the same as $\mathbf{a} \otimes \mathbf{a}$ after all. An example of why this doesn't violate the no-cloning theorem is that when $\mathbf{a}$ is a non-basis state, such as $\frac{1}{\sqrt{2}}(\mathbf{e}_x + \mathbf{e}_y)$, $\mathbf{a} \otimes \mathbf{a}$ is generally not the same as $\frac{1}{\sqrt{2}}(\mathbf{e}_{xx} + \mathbf{e}_{yy})$.

We can now do various things. We can run two operations $\boldsymbol{U}_f$ computing a function $f(x)$ on x side by side. Or we can do just one, applying $\boldsymbol{I}^{\otimes n} \otimes \boldsymbol{U}_f$ to $\boldsymbol{e}_{xx}$ to get $\boldsymbol{e}_{xf(x)}$. Essentially, we are using two Hilbert spaces that we put together by a product. We can also arrive at this kind of state in the manner shown next.

6.3 The Copy-Uncompute Trick

Suppose we wish to compute $f\colon \{0,1\}^n \to \{0,1\}^m$, where $m < n$. Such an f is not invertible, so we cannot expect to map an input state $\boldsymbol{e}_x$ to a quantum state that uniquely corresponds to y. We have already seen in sections 4.3 and 5.3 the idea of replacing f by the function

$$F(x,v) = (x, v \oplus f(x)).$$

Then $F\colon \{0,1\}^{n+m} \to \{0,1\}^{n+m}$ is a bijection, and the original function f is recoverable via $F(x,0^m) = (x,f(x))$.

Now suppose we have any quantum operation $\boldsymbol{U}$ on the "x" part, where $f(x)$ might be embedded as a substring in m indexed places. We can automatically obtain the corresponding $F(x)$ via the computation:

$$(\boldsymbol{U}^* \otimes \boldsymbol{I}_m)\boldsymbol{C}_m(\boldsymbol{U} \otimes \boldsymbol{I}_m)(\boldsymbol{e}_x \otimes \boldsymbol{e}_{0^m}),$$

where the $\boldsymbol{C}_m$ is applied to those index places and to m ancilla places. This effectively lifts out and copies $f(x)$ into the fresh places. The final $\boldsymbol{U}^*$ then inverts what $\boldsymbol{U}$ did in the first n places, "cleaning up" and leaving x again. Here is a diagram for $n = 4$ and $m = 2$ where the values $f(x) = y_1 y_2 \in \{0,1\}^2$ are computed on the second and third wires and then copied to the ancillae:

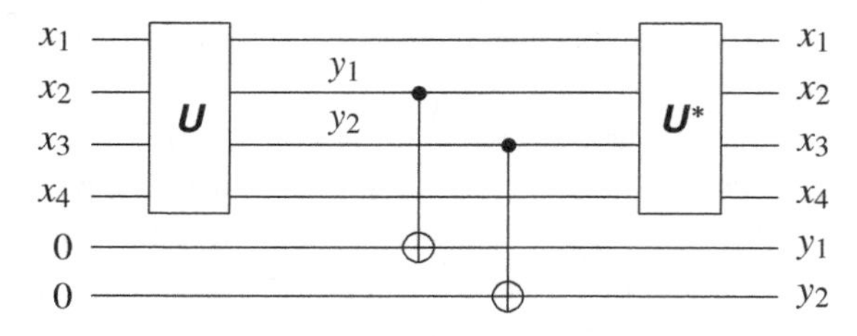

This trick is called **copy-uncompute** or **compute-uncompute**. It is important to note that it works only when the quantum state after applying $\boldsymbol{U}$ and before $\boldsymbol{C}_m$ is a superposition of only those basis states that have $f(x)$ in the set of quantum coordinates to which the controls are applied. If there is any disagreement there in the superposition, then the results can be different.

This again is why the trick does not violate the no-cloning theorem. For a simple concrete example, consider the following quantum circuit, noting that the Hadamard matrix is its own inverse, i.e., is self-adjoint:

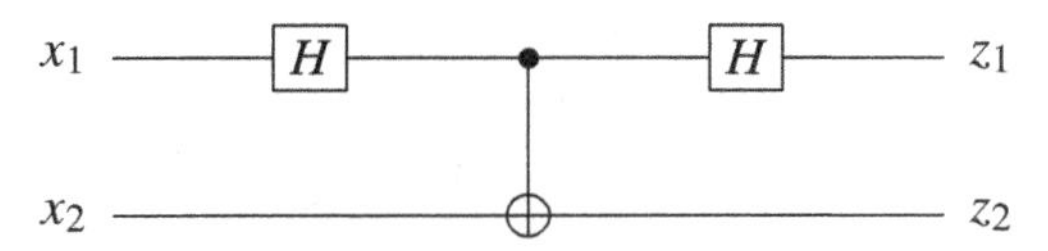

On input $\boldsymbol{e}_{00}$, that is, $x_1 = x_2 = 0$, the first Hadamard gate gives the control qubit a value that is a superposition. Hence, the second Hadamard gate does *not* "uncompute" the first Hadamard to restore $z_1 = 0$. The action can be worked out by the following matrix multiplication (with an initial factor of $\frac{1}{2}$):

$$\begin{bmatrix} 1 & 0 & 1 & 0 \\ 0 & 1 & 0 & 1 \\ 1 & 0 & -1 & 0 \\ 0 & 1 & 0 & -1 \end{bmatrix} \begin{bmatrix} 1 & 0 & 0 & 0 \\ 0 & 1 & 0 & 0 \\ 0 & 0 & 0 & 1 \\ 0 & 0 & 1 & 0 \end{bmatrix} \begin{bmatrix} 1 & 0 & 1 & 0 \\ 0 & 1 & 0 & 1 \\ 1 & 0 & -1 & 0 \\ 0 & 1 & 0 & -1 \end{bmatrix} = \begin{bmatrix} 1 & 1 & 1 & -1 \\ 1 & 1 & -1 & 1 \\ 1 & -1 & 1 & 1 \\ -1 & 1 & 1 & 1 \end{bmatrix}.$$

This maps $\boldsymbol{e}_{00}$ to $\frac{1}{2}[1, 1, 1, -1]$, thus giving equal probability to getting 0 or 1 on the first qubit line.

However, if $\boldsymbol{U}$ includes a preamble transforming $\boldsymbol{e}_0$ to $\boldsymbol{e}_x$ and then leaves a definite value y on the controlled lines before the rest of the circuit does $\boldsymbol{U}^*$, then the computation does end with the first n places again zeroed out, i.e., in some state $\boldsymbol{f}' = \boldsymbol{e}_{0^n} \otimes \boldsymbol{e}_y$. This finally justifies why we can regard $\boldsymbol{e}_0$ as the only input we need to consider. It emphasizes the goal of efficiently *preparing* a state from which a desired value $f(x)$ can be recovered by measurement.

As long as we are careful to represent the linear algebra correctly, we will not be confused between these two eventualities. Then we can do more tricks with superpositions and controls.

6.4 Superposition Tricks

Recall our $\boldsymbol{j}_N$ vector, which in the case $n = 2$, $N = 4$ is $\frac{1}{2}[1, 1, 1, 1]$. Feeding it on the first n of $2n$ quantum coordinates and following it with controls gives the following state:

$$(\boldsymbol{C}_n(\boldsymbol{j}_N \otimes \boldsymbol{e}_0))(xy) = \begin{cases} \frac{1}{\sqrt{N}} & \text{if } y = x \\ 0 & \text{otherwise.} \end{cases}$$

Furthermore, these ideas show that we can construct a vector $\boldsymbol{b}$ such that

$$\boldsymbol{b}(xy) = \begin{cases} \frac{1}{\sqrt{N}} & \text{when } y = f(x) \\ 0 & \text{otherwise.} \end{cases}$$

Here xy is just the concatenation of the strings x and y. Moreover, by the last section, we can obtain a version of $\boldsymbol{b}$ even when y is just a single bit. In either case, we can also write

$$\boldsymbol{b} = \frac{1}{\sqrt{N}} \sum_{x \in \{0,1\}^n} (\boldsymbol{e}_x \otimes \boldsymbol{e}_{f(x)}).$$

This is one of two instances where we find it most apt to mention Dirac notation.[1] The Dirac notation for this state is

$$\boldsymbol{b} = \frac{1}{\sqrt{N}} \sum_{x \in \{0,1\}^n} |x\rangle |f(x)\rangle.$$

DEFINITION 6.2 Given $f\colon \{0,1\}^n \to \{0,1\}^m$, the state $\boldsymbol{s}_f = \frac{1}{\sqrt{2^n}} \sum_x |x\rangle |f(x)\rangle$ is called the **functional superposition** of f.

We can also extend the conditional idea of $\boldsymbol{C_n}$ directly to any given quantum operation $\boldsymbol{U}$. Define $\boldsymbol{CU}$ by

$$((\boldsymbol{CU})\boldsymbol{a})(0x) = \boldsymbol{a}(x); \quad ((\boldsymbol{CU})\boldsymbol{a})(1x) = (\boldsymbol{Ua})(x).$$

We have used extra parentheses to make clear that $\boldsymbol{CU}$ is a name, not the composition of matrices called $\boldsymbol{C}$ and $\boldsymbol{U}$, and it is read "Control-$\boldsymbol{U}$." Our $\boldsymbol{CNOT}$ operation did this to our matrix $\boldsymbol{X}$ of the unitary NOT operation, which explains the name. We can also iterate this, for instance, to do $\boldsymbol{CCNOT}$. This yields our friend the Toffoli gate again.

6.5 Flipping a Switch

There are many old jokes of the form, How many X-es does it take to change a light bulb? In quantum computation, everything is reversible, and that applies

1 The other involves writing the projector $\boldsymbol{P_a}$ defined in section 5.5 via the **outer product** matrix, which is defined generally for all vectors $\boldsymbol{a}, \boldsymbol{b}$ by $|\boldsymbol{a}\rangle\langle\boldsymbol{b}|[i,j] = \boldsymbol{a}(i)\overline{\boldsymbol{b}(j)}$, as $\boldsymbol{P_a} = |\boldsymbol{a}\rangle\langle\boldsymbol{a}|$. Thus, for all vectors $\boldsymbol{x}$, $\boldsymbol{P_a}x = |\boldsymbol{a}\rangle\langle\boldsymbol{a}|x = \boldsymbol{a}\langle\boldsymbol{a},\boldsymbol{x}\rangle$. While $\boldsymbol{P_a}$ is not unitary, it contributes to the unitary reflection operator $\boldsymbol{Ref_a} = 2\boldsymbol{P_a} - \boldsymbol{I}$.

to jokes as well: If you change a light bulb, how many X-es can you affect? The answer is: as many as you like.

Our light bulb can be the $(n+1)$st qubit, call it y. Suppose we multiply it by a unit complex number a, such as -1. It may seem that we are only flipping the sign of the last qubit, and we might even wrongly picture the $(n+1) \times (n+1)$ matrix that is the identity except for a in the bottom right corner. The unitary matrices that are really involved, however, are $2^{n+1} \times 2^{n+1}$ acting on the Hilbert space, and by linearity, the scalar multiplication applies to *all* coordinates. Put another way, if we start with a product state $\boldsymbol{z} \otimes \boldsymbol{e}_y$ and change the latter part to $a\boldsymbol{e}_y$, then the resulting tensor product is mathematically the same as $(a\boldsymbol{z}) \otimes \boldsymbol{e}_y$. With $a = -1$, we can interpret this as $\boldsymbol{z}$ being flipped instead. This feels strange, but both come out the same in the index-based calculations.

This becomes a great trick if we can arrange for a itself to depend on the basis elements $\boldsymbol{e}_x$. Given a Boolean function f with one output bit, let us return to the computation of the reversible function $F(x,y) = (x, (y \oplus f(x)))$. Our quantum circuits for f have thus far initialized y to 0. Let us instead arrange $y = 1$ and then apply a single-qubit Hadamard gate. Thus, instead of starting up with $\boldsymbol{e}_x \otimes \boldsymbol{e}_0$, we have $\boldsymbol{e}_x \otimes \boldsymbol{d}$, where $\boldsymbol{d}$ is the "difference state"

$$\boldsymbol{d} = (\frac{1}{\sqrt{2}}, \frac{-1}{\sqrt{2}}) = \frac{1}{\sqrt{2}}(\boldsymbol{e}_0 - \boldsymbol{e}_1).$$

Now apply the circuit computing F. By linearity we get:

$$\begin{aligned} F(x,d) &= \frac{1}{\sqrt{2}}(F(x0) - F(x1)) \\ &= \frac{1}{\sqrt{2}}\left(\boldsymbol{e}_x \otimes \boldsymbol{e}_{0\oplus f(x)} - \boldsymbol{e}_x \otimes \boldsymbol{e}_{1\oplus f(x)}\right) \\ &= \frac{1}{\sqrt{2}}\left(\boldsymbol{e}_x \otimes (\boldsymbol{e}_{0\oplus f(x)} - \boldsymbol{e}_{1\oplus f(x)})\right) \\ &= \boldsymbol{e}_x \otimes \boldsymbol{d}', \end{aligned}$$

where

$$\begin{aligned} \boldsymbol{d}' &= \begin{cases} \frac{1}{\sqrt{2}}(\boldsymbol{e}_0 - \boldsymbol{e}_1) & \text{if } f(x) = 0 \\ \frac{1}{\sqrt{2}}(\boldsymbol{e}_1 - \boldsymbol{e}_0) & \text{if } f(x) = 1 \end{cases} \\ &= (-1)^{f(x)}\boldsymbol{d}. \end{aligned}$$

Thus, we have flipped the last quantum coordinate by the value $a_x = (-1)^{f(x)}$. Well actually no—by the above reasoning, what we have equally well done is that when presented with a basis vector $\mathbf{e}_x$ as input, we have multiplied it by the x-dependent value a_x. We have involved the $(n+1)st$ coordinate, but because we have obtained $a_x\mathbf{e}_x \otimes \mathbf{d}$, we can regard it as unchanged. In fact, we can finish with another Hadamard and ***NOT*** gate on the last coordinate to restore it to 0. On the first n qubits, over their basis vectors $\mathbf{e}_x$, what we have obtained is the action

$$\mathbf{e}_x \mapsto (-1)^{f(x)}\mathbf{e}_x.$$

This is the action of the Grover oracle. We have thus proved theorem 5.6 in chapter 5. We can summarize this and the conclusion of section 6.4 in one theorem statement:

THEOREM 6.3 For all (families of) functions $f\colon \{0,1\}^n \to \{0,1\}^m$ that are classically feasible, the mapping from $\mathbf{e}_{x0^m}$ to the functional superposition $\mathbf{s}_f$ and the Grover oracle of f are feasible quantum operations. □

6.6 Measurement Tricks

There are also several tricks involving measurement. Suppose that the final state of some quantum algorithm is $\mathbf{a}$. We now plan to take a measurement that will return y with probability $|\mathbf{a}(y)|^2$. In some cases, we can compute this in closed form, whereas in other cases, we can approximate it well. In other algorithms, we use the following idea: "Everybody has to be somewhere."

Let S be a subset of the possible indices y, and suppose that we can prove

$$\sum_{y \in S} |\mathbf{a}(y)|^2 \geq c > 0$$

for some constant c. Then we can assert that with probability at least c a measurement will yield a good y from the set S. Note, the power of this trick is that we do not have to understand the values of $\mathbf{a}^2(z)$ for z *not* in the set S. We need only understand those in the set S. This is used in chapter 11 when we study Shor's factoring algorithm.

When the set S is the set of all indices having a "1" in a particular place, this is called *measuring one qubit*. Note that S includes exactly half of the indices, as does its complement, which equally well defines a one-qubit measurement. The idea can be continued to define r-qubit measurements, each of which "targets" a particular outcome string $w \in \{0,1\}^r$ and involves the particular set S_r of $N/2^r$ indices that have w in the respective places.

Theoretically, after a one-place measurement, the quantum computation can continue on the smaller Hilbert space of the remaining places. This does not matter to us because all algorithms we cover do their measurements at the end of quantum routines. As a footnote, we cover the **principle of deferred measurement**, which often removes the need to worry about this possibility because it illustrates the above controlled-$\boldsymbol{U}$ trick. We state it as a theorem.

THEOREM 6.4 If the result b of a one-place measurement is used only as the test in one or more operations of the form "if b then $\boldsymbol{U}$," then exactly the same outputs are obtained upon replacing $\boldsymbol{U}$ by the quantum controlled operation $\boldsymbol{CU}$ with control index the same as the index place being measured and measuring that place later without using the output for control.

Proof. As before, we visualize the control index being the first index, but it can be anywhere. Suppose in the new circuit the result of the measurement is 0. Then the $\boldsymbol{CU}$ acted as the identity, so on the first index, the same measurement in the old circuit would yield 0, thus failing the test to apply $\boldsymbol{U}$ and so yielding the identity action on the remainder as well. If the new circuit measures 1, then because $\boldsymbol{CU}$ does not affect the index, the old circuit measured 1 as well, and in both cases the action of $\boldsymbol{U}$ is applied on the remainder. □

6.7 Partial Transforms

Another trick is applying an operator to "part" of the space. The general principle is really quite simple. Suppose that $\boldsymbol{U}$ is a unitary transform defined on vectors $\boldsymbol{a}$ in some Hilbert space. Then by definition there is a function $u(x,k)$ so that

$$(\boldsymbol{Ua})(x) = \sum_k u(x,k)\boldsymbol{a}(k).$$

Now suppose that we want to extend $\boldsymbol{U}$ to apply to the vector $\boldsymbol{b}(xy)$. The question is how do we do this? The natural idea is to imagine that $\boldsymbol{b}$ is really many different vectors, each of the form $\boldsymbol{b}(xy_0)$ for a different fixed value of y_0. If we want the result of applying $\boldsymbol{U}$ to one of these, it should be

$$\sum_k u(x,k)\boldsymbol{b}(xy_0).$$

Therefore, the result of applying $\boldsymbol{U}$ to the vector $\boldsymbol{b}$ is:

$$\boldsymbol{c}(xy) = \sum_k u(x,k)\boldsymbol{b}(k,y).$$

As an example, let $\boldsymbol{a}(xy)$ be a vector. We can apply the Hadamard transform just to the first "x" part as follows. The result is

$$\boldsymbol{b}(xy) = \frac{1}{\sqrt{N}} \sum_t (-1)^{x \bullet t} \boldsymbol{a}(ty).$$

This is what is meant by applying the Hadamard only to the "x" coordinates.

6.8 Problems

6.1. Suppose that $\boldsymbol{a}$ is a unit vector. If we know that $\boldsymbol{a}(k) = 1$, then what can we say about $\boldsymbol{a}(\ell)$ for $\ell \neq k$?

6.2. Fix a dimension N. Consider the group of unitary matrices generated by the Hadamard matrices $\boldsymbol{H}_M$ for $M = 2^m$, $m = 1,\dots,n = \log_2 N$ and all permutation matrices that swap two qubit lines. What are the sizes of the group for $N = 2$ and $N = 4$? What about for general $N > 4$?

6.3. Consider all unit vectors of dimension 4 that have $\frac{\pm 1}{2}$ as entries. Show that all can be obtained from one another by using only the transformations in the previous problem.

6.4. Prove the no-cloning theorem in the case $n = 2$: Suppose for sake of contradiction that $\boldsymbol{U}$ is a unitary matrix such that for any $\boldsymbol{a} = [a,b]^T$,

$$\boldsymbol{U}(\boldsymbol{a} \otimes \boldsymbol{e}_0) = \boldsymbol{a} \otimes \boldsymbol{a}.$$

To reach a contradiction, take a second arbitrary state $\boldsymbol{b}$ and use the property that unitary matrices preserve inner products.

6.5. Show how to construct a unitary matrix $\boldsymbol{U}$ such that $\boldsymbol{U}[a,b,0,0]^T = \frac{1}{\sqrt{2}}[a,b,a,b]^T$. Why doesn't this contradict the no-cloning theorem?

6.6. Recall the real-valued rotation matrices $\boldsymbol{R_x}(\theta)$ from problems 3.15–3.16 in chapter 3. Show that the controlled rotation $\boldsymbol{CR_x}(\theta)$ can be simulated by two $\boldsymbol{CNOT}$ gates sandwiched in with the half-rotation $\boldsymbol{R_x}(\theta/2)$ and its inverse $\boldsymbol{R_x}(-\theta/2)$ on the target qubit line.

6.7. Recall the $\boldsymbol{T}$ and $\boldsymbol{V}$ matrices from exercises in chapter 3:

$$\boldsymbol{T} = \begin{bmatrix} 1 & 0 \\ 0 & e^{i\pi/4} \end{bmatrix}, \quad \boldsymbol{V} = \frac{1}{\sqrt{2}} \begin{bmatrix} e^{i\pi/4} & e^{-i\pi/4} \\ e^{-i\pi/4} & e^{i\pi/4} \end{bmatrix}.$$

Write out the matrices of their controlled versions, $\boldsymbol{CT}$ and $\boldsymbol{CV}$. What are their squares?

6.8. Show that $\boldsymbol{CV}$ can be written as a composition of (appropriate powers of) $\boldsymbol{T}$ gates before and after a controlled rotation $\boldsymbol{CR_x}(\theta)$. Deduce that $\boldsymbol{CNOT}$ is the only two-qubit gate needed to simulate $\boldsymbol{CV}$.

The next two problems contribute to the converse direction of simulating the $\boldsymbol{T}$-gate via Hadamard and Toffoli, which will be finished in the next chapter's exercises. Recall problem 3.12 from chapter 3, where we computed the *group commutator* of the 2×2 matrices $\boldsymbol{H}$ and $\boldsymbol{S}$.

6.9. Compute the commutator of $\boldsymbol{I} \otimes \boldsymbol{H}$ with

$$\boldsymbol{CS} = \begin{bmatrix} 1 & 0 & 0 & 0 \\ 0 & 1 & 0 & 0 \\ 0 & 0 & 1 & 0 \\ 0 & 0 & 0 & i \end{bmatrix}.$$

Now is it possible to multiply by a scalar c such that all entries are powers of i?

6.10. Undo the last part of the commutator in the last problem—that is, sandwich $\boldsymbol{CS}$ between two Hadamard gates on the second qubit line. What gate do you have? Conclude that Hadamard and $\boldsymbol{CS}$ suffice to simulate the $\boldsymbol{CV}$ gate using one extra qubit line.

The last problems here show how to use controlled gates to decompose the quantum Fourier transform into basic gates.

6.11. Show that for any $2^k \times 2^k$ unitary matrix $\boldsymbol{A}$, the block matrix

$$\begin{bmatrix} \boldsymbol{I}^{\otimes k} & \boldsymbol{A} \\ \boldsymbol{I}^{\otimes k} & -\boldsymbol{A} \end{bmatrix}$$

can be written as the composition of $\boldsymbol{H} \otimes \boldsymbol{I}^{\otimes k}$ and the controlled matrix $\boldsymbol{CA}$.

6.12. If $\boldsymbol{A}$ in problem 6.11 is a diagonal matrix, how many two-qubit controlled gates do you need to simulate $\boldsymbol{CA}$?

6.13. Recalling the recursive equation for the quantum Fourier transform in problem 5.6,

$$\mathbf{F}_N = \frac{1}{\sqrt{2}} \begin{bmatrix} \mathbf{I}^{\otimes(n-1)} & \mathbf{D}_{N/2} \\ \mathbf{I}^{\otimes(n-1)} & -\mathbf{D}_{N/2} \end{bmatrix} \begin{bmatrix} \mathbf{F}_{N/2} & 0 \\ 0 & \mathbf{F}_{N/2} \end{bmatrix} \mathbf{K}_n,$$

apply problems 6.11 and 6.12 to express it all as a composition of Hadamard gates, swap gates, and 2-qubit controlled twists. Mindful that the second matrix here is just $\mathbf{F}_{N/2} \otimes \mathbf{I}$, how many gates do you need?

6.9 Summary and Notes

As in other areas, the key to understanding is often more than knowing the "big" results—it is also important to know the important little tricks. This chapter has exemplified some tricks using our index-based notation for vectors and matrices. The tricks focus on the parts of linear algebra that are most relevant to quantum computing, so for further reading we suggest texts such as Nielsen and Chuang (2000), Yanofsky and Mannucci (2008), Hirvensalo (2010), and Rieffel and Polak (2011).

We have attempted to distinguish between tricks of general algorithmic importance and tricks and properties of specific quantum gates. We have put a lot of the latter in the exercises of chapters 3, 5, and here. The text by Williams (2011) has a cornucopia of further details about quantum gates, and even more can be said about engineering issues for gates and circuits. We have regarded these exercises first as giving practice in linear algebra, and second as a reasonable substitute for in-text coverage of model-specific simulation theorems. The message of the last problems here is that commutators involving Hadamard and θ-angled phase gates, with **CNOT** and ancilla qubits to nail things down, enable simulating the effect of gates of phase angles $\theta/2$, $\theta/4$, $\theta/8$, and so on. We have chosen not to go further in proving that this enables sufficiently fine approximation of the twists $\mathbf{T}_{\pi/2^{n-1}}$ involved in the representation of the QFT in problems 6.13, as the full details would take us out of scope now. This formally justifies counting the QFT as *feasible* with these gate sets, in Shor's algorithm and other applications, but we prefer to take the entire QFT as basic while presenting the algorithms.

7 Phil's Algorithm

There is no quantum algorithm named after Phil—at least none that we know about. The goal of this chapter is to give the schema we will use for presenting all the rest of the quantum algorithms and say how they give their results. We will then tell you who Phil is.

We will always start with a description of what the algorithm actually does. This will usually be of the form:

> Given an X, Phil's algorithm finds a Y within time Z.

Sometimes the goal is achieved always, otherwise it comes with a specified probability or expected length of time. We may add some additional comments on why the problem is interesting and important.

7.1 The Algorithm

Each algorithm will be presented as computing a series of vectors. In the first few algorithms, the number of vectors is fixed independent of the size of the input object X. This is an interesting point because you might have expected that the number of vectors would grow as X gets bigger. The reason for this is that each vector corresponds to a macrophase of the algorithm. Often the algorithms have only a small number of macrophases, where each phase does something different.

In describing the vectors, we will always explain what Hilbert space they are from. Again, there is some commonality: all but those using the quantum Fourier transform are directly understandable in real spaces, whereas Shor's algorithm uses a complex Hilbert space. Problems 7.8–7.14 explore this matter further.

7.2 The Analysis

Quantum algorithms are similar to classical ones in that often the algorithms' descriptions are simpler than their analysis. The analysis of these algorithms usually will consist of giving an explicit description of what the i-th vector is. The algorithm gives the operational description of the vector: it is the result of applying unitary transformations to the start vector. Here we give a non-operational description: the vector is described by the following mathematical expression.

Once we know what each vector is explicitly, we will know what the last vector is. Then we will understand what the result of the last step of each algorithm does because in all cases the last step is a quantum measurement. Of course to understand measurements, we must know the amplitudes given by the last vector because the measurement returns k with the amplitude squared of the k-th coordinate.

There is one more complication—isn't there always? Some algorithms are finished after the measurement is made: the measurement's value determines the answer completely. Other algorithms require that additional classical processing is performed on the result of the measurement. Some are a bit more involved, in that they need the quantum algorithm to be run a multiple number of times. Each run of the quantum algorithm gives a small amount of information about the answer that is desired. This happens, for example, with both Simon's and Shor's algorithms.

7.3 An Example

Here is Phil's algorithm—we said there was no such algorithm—so we made one up. It operates over a two-dimensional Hilbert space $\mathbb{H}_2$. The start vector $\boldsymbol{a}_0$ is

$$\begin{bmatrix} 1 \\ 0 \end{bmatrix}.$$

The next vector is $\boldsymbol{a}_1$, which is equal to $\boldsymbol{H}_2\boldsymbol{a}_0$—of course $\boldsymbol{H}_2$ is the 2×2 Hadamard transform. Then we measure this vector and return the index 0 or 1. We see 0 with probability $\boldsymbol{a}_1^2(0)$ and 1 with probability $\boldsymbol{a}_1^2(1)$. That is it—not too exciting, but it does compute something. What it provides is the ability to flip a fair coin. This will be a building block of other algorithms.

7.4 A Two-Qubit Example

Phil becomes a bit more ambitious now, so he has two qubits. He carries out the composition of $\boldsymbol{V_1} = \boldsymbol{H} \otimes \boldsymbol{I}$ and $\boldsymbol{V_2} = \boldsymbol{CNOT}$, which we illustrated in chapter 4. We index vectors in this two-qubit space by xy, where x and y are single bits. In the itemized format we use for quantum algorithms, this is what he does:

The Algorithm

1. The initial vector is $\boldsymbol{a}_0$ so that $\boldsymbol{a}_0(00) = 1$—that is, $\boldsymbol{a}_0 = \boldsymbol{e}_{00}$.
2. The next vector $\boldsymbol{a}_1$ is the result of applying the Hadamard transform on qubit line 1 only.
3. The final vector $\boldsymbol{a}_2$ is the result of applying ***CNOT*** to $\boldsymbol{a}_1$.

The Analysis

First,

$$\boldsymbol{a}_1 = \frac{1}{\sqrt{2}}(\boldsymbol{e}_0 + \boldsymbol{e}_1) \otimes \boldsymbol{e}_0 = \frac{1}{\sqrt{2}}(\boldsymbol{e}_{00} + \boldsymbol{e}_{10}) = \frac{1}{\sqrt{2}}[1,0,1,0].$$

Then because ***CNOT*** swaps the third and fourth Hilbert-space coordinates, we can jump right away to see that

$$\boldsymbol{a}_2 = \frac{1}{\sqrt{2}}[1,0,0,1].$$

Thus far, we have not said anything about taking measurements—instead, we are able to specify the final pure quantum state we get. In the Hilbert space coordinates it doesn't look exciting, but let's interpret it back in the quantum coordinates:

$$\boldsymbol{a}_2 = \frac{1}{\sqrt{2}}(\boldsymbol{e}_{00} + \boldsymbol{e}_{11}).$$

This state is pure and not a tensor product of two other states, so it is entangled. This matters immediately if and when we do a measurement. If we measure both qubits, then we will only get 00 or 11, never 01 or 10. If we measure just the first quantum coordinate and get 0, then we know already that any measurement of the second quantum coordinate will give 0. Thus, Phil's algorithm has produced an entangled pair of qubits.

This becomes significant when we are able to give the first qubit to someone named "Alice" sitting 10 miles east of Lake Geneva and the second qubit to her friend "Bob" sitting 10 miles west, and each does a measurement at instants such that no signal of Alice's result can reach Bob before he measures and vice versa. Whatever result Alice gets, Bob gets too. Albert Einstein called the effect "spooky" because it appeared to violate his own established principle that influence could not propagate faster than light, but that worry hasn't stopped real Alices and real Bobs from executing this algorithm at a distance. So thinking in physical terms, Phil's little algorithm was good enough to stump the great Einstein. However, we do not have to think in physical terms—Phil's output is just an ordinary vector in our four-dimensional Hilbert space. What we do need to think more about, to finish the analysis of our algorithms, is measurement.

7.5 Phil Measures Up

Now we will tell you who Phil is. Phil is a mouse. Unlike a certain famous quantum cat, who does nothing except lie around half-dead all day, Phil is very active. Phil runs through mazes like many other laboratory mice, but there are some special things about Phil:

- Phil runs through every path in the maze at once. Like we said, he is very active. He follows Yogi Berra's advice: when he comes to a fork in the road, he takes it, becoming two Phils.
- Some corridors in the mazes have a piece of cheese. When Phil eats a piece of cheese, he turns into Anti-Phil. If Phil runs into Anti-Phil, they annihilate each other, leaving *nothing*. Not a combination, as with Schrödinger's cat, but really nothing. However, if Anti-Phil eats a second piece of cheese, then he turns into Phil again. A third piece makes Anti-Phil again, and so on.
- If Phil meets himself—not Anti-Phil—where corridors come together, then they run alongside each other. If Anti-Phil meets Anti-Phil, then they likewise start to form an Anti-Phil pack. If two opposite packs meet, then they still cancel each other in pairs, leaving just the surviving members of one pack or the other, if any.
- If a pack reaches an exit of the maze safely, then it can be put under an incubator that mutantly grows it to the square of the number in the pack. The mutant is the same whether the pack has Phils or Anti-Phils—it is a "Mighty Mouse." Dividing its size by a certain number that depends on the number of stages with cheese and the manner of measurement gives a value between 0 and 1, which is the probability that Phil exits the maze there.
- After the division, the mutants "collapse," and Phil comes together as just one ordinary mouse again. Or we think he does—the chapter end notes briefly mention some debate about this. At least we can say that no laboratory animals were harmed in the course of writing this book.

Phil can eat several kinds of cheese, but one French type is far and away his favorite: Hadamard cheese. A corridor with Hadamard cheese is labeled -1 in the maze blueprints. More complex types of cheese might have labels like i and $-i$ and $e^{i\pi/4}$ and $e^{3i\pi/4}$, and these would create "Half-Moon Phil" and "Crescent Phil" and "Gibbous Phil" and other "phased-out" spectral mice. But as we will state formally in chapter 16, Hadamard cheese and "Anti-Phil" are ultimately a rich enough basis. Assuming that the only nondeterministic gates are h-many Hadamard gates and all mice are measured, the division number is 2^h.

The mazes have N entrances, one for every basis vector $\mathbf{e}_x$, and N exits with the same labels. They are built in stages, each with N opening and closing junctures keeping the same labels, one stage for each basic quantum operation. A Hadamard stage gives a choice of two corridors: one going straight across and one diagonally. Cheese is placed in half of the horizontal corridors—those whose label has a "1" in a certain one of the n places. The other stages we need to consider are all permutations of $\{0, 1\}^n$ and are built by routing the corridors according to the permutation, giving no other choice. Corridors may "cross" each other—that is, the maze is three-dimensional. Figure 7.1 gives the maze for the above computation with one Hadamard gate and one ***CNOT*** gate.

Figure 7.1
Maze for Hadamard on qubit 1 followed by ***CNOT*** on 1 and 2.

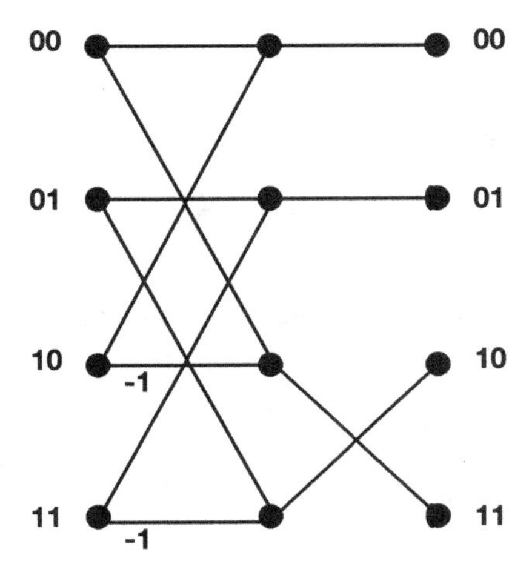

To interpret this, suppose Phil enters the maze at the upper left, that is, with input 00. He can run across or down. Well, he does both. One Phil ends up at 00, the other at 11. Although neither got to eat any cheese, the presence of one Hadamard stage says to divide by 2 after squaring. Thus, each of the two outcomes has measure $1^2/2^1 = 0.5$. The outcomes 01 and 10 have no Phils, so they measure 0.

If Phil starts running at entrance 01, then each of those outcomes gets one Phil while 00 and 11 get none.

Anti-Phil makes an appearance if we start Phil at 10. Phil scampers up to 00 and stays there without eating cheese, but another Phil scampers across, eats cheese, and in the second stage jumps down to 11 as Anti-Phil. Thus, the final state is different from the case before: allowing for the cheese factor, it is $\frac{1}{\sqrt{2}}[1,0,0,-1]$ rather than $\frac{1}{\sqrt{2}}[1,0,0,1]$. But its measurements are the same: it gives probability 0.5 on 00 and 0.5 on 11. People who bet on where the mouse ends up don't care—Anti-Phil gives the same measurement value as Phil. On input 11, we similarly get Phil at 01 and Anti-Phil at 10.

For the simplest case where Phil and Anti-Phil collide, see figure 7.2, which shows two consecutive Hadamard gates on a single qubit. We have annotated each juncture with the "Phil counts" for the respective pair of entrances 0 and 1, where positive values denote a Phil pack and negative values an Anti-Phil pack.

Figure 7.2
Maze for two consecutive Hadamard gates.

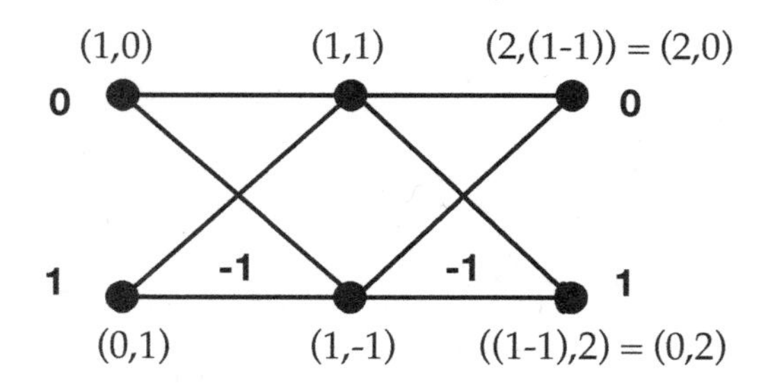

Again, suppose Phil starts at the upper left, from 0. One Phil scampers straight across twice to exit at 0, while another scampers diagonally down to 1 and back up to meet him at that exit. They exist as a pack of 2, which gets squared to 4. Because there are $h = 2$ Hadamards, this gets divided by $2^h = 4$. Thus, the probability of exiting at 0 is 1.0, which doesn't leave room for exiting anywhere else. But what about the other Phils?

The first Phil who scampered straight ahead has a second choice and can scamper diagonally down to the exit 1, still eating no cheese and hence ending there as Phil. The second Phil who scampered down, however, can stay down to eat the cheese and ends up at 1 as Anti-Phil. The Phil and Anti-Phil at 1 cancel, leaving 0 as their measurement. Thus, it is impossible for Phil entering at 0 to exit at 1—people who bet on that outcome will never win.

Similarly, if Phil enters at 1, then he must exit at 1. But note what happened: one Phil scampered up and then back down and never ate any cheese. The other scampered across to be Anti-Phil, but then ate the other cheese to become Phil again just in time. Thus, the two consecutive Hadamard gates leave the same condition on exit as they had on entering—they are equivalent to the identity operation, which means just corridors going straight across with no choices. Whether they are really the same as "no-operation" is another matter for debate—in practice, it seems that the real Phils get some wear and tear and mussed-up hair and give measurements that are close to 1.0 and 0.0 or to other intended values like 0.5 but not quite equal.

Of course this is just an elaboration of what happens when you multiply the matrices. Here we are just doing

$$\boldsymbol{H} \cdot \boldsymbol{H} = \frac{1}{\sqrt{2}} \begin{bmatrix} 1 & 1 \\ 1 & -1 \end{bmatrix} \frac{1}{\sqrt{2}} \begin{bmatrix} 1 & 1 \\ 1 & -1 \end{bmatrix} = \frac{1}{2} \begin{bmatrix} 2 & 0 \\ 0 & 2 \end{bmatrix} = \boldsymbol{I}.$$

Indeed, every self-adjoint matrix $\boldsymbol{U}$, meaning $\boldsymbol{U} = \boldsymbol{U}^*$, that is also unitary cancels itself out this way when squared. Using "Phil" helps visualize the quantum-mechanical effects of *superposition* and *interference*, but the math is just linear algebra. Hence, after the next chapter, we will "retire" him and do proofs formally, but he will make a useful return in the last chapters.

7.6 Quantum Mazes versus Circuits versus Matrices

The main problem with our maze diagrams is that they do not *scale*—they grow with N, which is exponential in n. Hence, most sources in quantum computation prefer to write circuit diagrams, which scale with n instead. Here again is our diagram for the Hadamard plus ***CNOT*** circuit, except that now we wish to label the values carried by the circuit's individual "wires":.

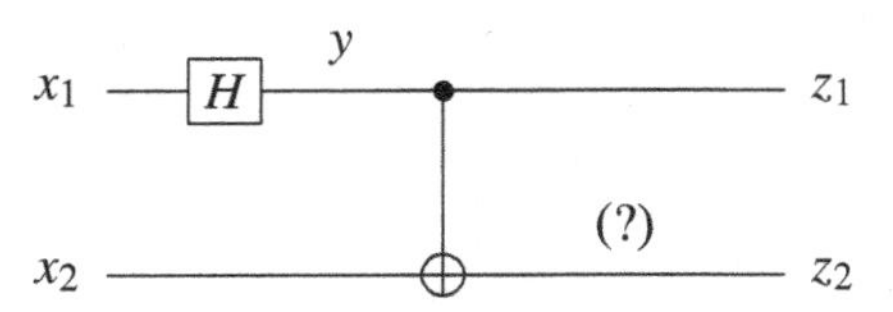

Here x_1 and x_2 are variables denoting the input qubits, z_1 and z_2 are the outputs, and y is a variable denoting the choice offered at the Hadamard gate. Our unease with the circuit diagram, however, is represented by the "(?)" label:

Whereas a Boolean circuit, on a given input, always has a definite value (0 or 1) at every gate juncture, this is not always true of a quantum circuit owing to entanglement.

The value at "(?)" could be 0 or 1, with equal probability in fact (and hence equal amplitudes $1/\sqrt{2}$), but it is not even correct to say that $(\boldsymbol{e}_0 + \boldsymbol{e}_1)/\sqrt{2}$ is its value. The value is *entangled* with the value of y, with both depending on the input values.

The similar diagram for two Hadamard gates is even simpler but does not immediately help us calculate that they cancel:

x — H — y — H — z

Moreover, we expressed in section 6.5 the opinion that showing a scalar multiplication as occurring *on* a qubit line can be misleading, and this extends to other kinds of *phase* transformations.

The problem is that the mazes do not scale, whereas the circuits scale but make entanglements and some other information hard to trace. The advantage of our functional notation for vectors and matrices is that it scales while preserving everything. However, as experienced with functional programming notation in ordinary classical computing, it gives less "feel" for the objects. Hence, we will still often write out vectors and matrices in full.

The mazes are exactly the directed graphs corresponding to matrix products that were defined at the end of chapter 3, except that the adjacency matrices for the graphs of Hadamard stages can have -1 in place of $+1$. The mouse enters a matrix $\boldsymbol{U}$ in some row i and may exit in some column j if $\boldsymbol{U}[i,j] \neq 0$; if $\boldsymbol{U}[i,j] = -1$, then it picks up some cheese. Thus our mouse executes the product of the matrices, which is expressly visualized as a sum over paths in the graphs. This was the intuition of Richard Feynman originally with regard to so-called *S-matrices* representing dynamical systems. To animate his sum-over-paths formalism, we have named the mouse for his middle name, Phillips. That his middle name was plural makes it work even better.

7.7 Problems

7.1. Show that the vector $\boldsymbol{a}_1$ in section 7.3 is equal to

$$\frac{1}{\sqrt{2}}\begin{bmatrix}1\\1\end{bmatrix}.$$

Also, what does the algorithm actually do? Can you replace it by a classical one?

7.2. Suppose that there is a unitary matrix $\boldsymbol{U}$ so that $\boldsymbol{a} = \boldsymbol{U}\boldsymbol{e}_0$ where $\boldsymbol{a}(x) = 1/M$ if and only if x is a prime number in the range $0,\ldots,N-1$. What is the value of M? What happens when we perform a measurement on $\boldsymbol{a}$?

7.3. Draw the maze stage for a Toffoli gate using eight "levels" labeled 000 through 111. Suppose the Toffoli gate is upside-down, that is, it can alter the first rather than the third quantum coordinate. Then what does the maze stage look like?

7.4. Recalling the controlled-$\boldsymbol{V}$ operation from the last chapter's exercises,

$$\boldsymbol{CV} = \frac{1}{2}\begin{bmatrix}2 & 0 & 0 & 0\\ 0 & 2 & 0 & 0\\ 0 & 0 & 1+i & 1-i\\ 0 & 0 & 1-i & 1+i\end{bmatrix},$$

note how we can characterize it operationally:

$$\begin{aligned}\boldsymbol{CV}[0a,bc] &= \boldsymbol{I}[b,0]\boldsymbol{I}[a,c]\\ \boldsymbol{CV}[1a,bc] &= \boldsymbol{I}[b,1]\boldsymbol{V}[a,c].\end{aligned}$$

Now let $\boldsymbol{CIV}$ stand for the same idea with the "control" on the first qubit, but with the conditional $\boldsymbol{V}$ on the third qubit, while the second qubit is just ignored. By oddity of notation, we cannot write this as a tensor product of the

2×2 identity matrix $\boldsymbol{I}$ with $\boldsymbol{CV}$, as we could if the ignored qubit were the first or third. But we can describe it equally well operationally by

$$\begin{aligned} \boldsymbol{CIV}[0ab, cde] &= \boldsymbol{I}[0,c]\boldsymbol{I}[a,d]\boldsymbol{I}[b,e], \\ \boldsymbol{CIV}[1ab, cde] &= \boldsymbol{I}[1,c]\boldsymbol{I}[a,d]\,\boldsymbol{V}[b,e]. \end{aligned}$$

Write out the 8×8 matrix for this operation.

7.5. Show that the Toffoli gate $\boldsymbol{TOF}$ obeys the equation

$$\boldsymbol{TOF} = (\boldsymbol{I} \otimes \boldsymbol{CV})(\boldsymbol{CNOT} \otimes \boldsymbol{I})(\boldsymbol{I} \otimes \boldsymbol{CV}^*)(\boldsymbol{CNOT} \otimes \boldsymbol{I})\boldsymbol{CIV}.$$

Thus, the Toffoli gate can be written as a composition of two-qubit gates. Here is the famous quantum circuit diagram for this formula:

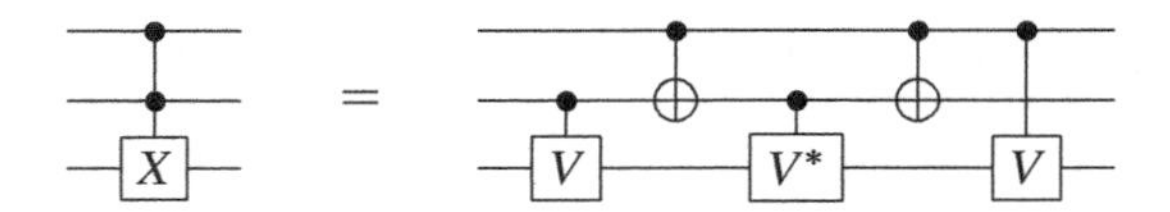

The intuition is that if the top bit is false, then the two $\boldsymbol{CNOT}$s and the final $\boldsymbol{CV}$ go away, leaving $\boldsymbol{CV}$ and $\boldsymbol{CV}^*$, which cancel, so the whole thing acts as the identity. If the top bit is true but the middle bit is false, then the first $\boldsymbol{CNOT}$ makes the middle bit true in time to activate the $\boldsymbol{CV}^*$, which then cancels with the second $\boldsymbol{CV}$ activated by the first bit. If both bits are true, then both $\boldsymbol{CV}$ gates are activated, whereas the middle bit becomes false and inhibits the $\boldsymbol{CV}^*$. This gives the action of $\boldsymbol{V}^2$, which equals $\boldsymbol{X}$. Thus, the whole action is $\boldsymbol{CCX}$, which equals the Toffoli gate.

The problem posed here is to verify the equation using our matrix indexing notation instead and then write a 5,000-word essay comparing it to the circuit intuition. OK, we are kidding about the 5,000-word part.

7.6. Deduce that the Toffoli gate can be simulated using $\boldsymbol{CNOT}$ and single-qubit matrices without needing any ancilla qubit lines (see also problems 6.6–6.8 of chapter 6).

7.7. Use the swap gate to write $\boldsymbol{CIV}$ as a composition of matrices, each of which is a tensor product of $\boldsymbol{I}$ with a 4×4 matrix. Conclude that $\boldsymbol{TOF}$ equals a composition with each term a tensor product of $\boldsymbol{I}$ and a 4×4 matrix.

The next problems complete a cycle of proving the equivalence of three famous gate sets as a basis for quantum computation and prove that matrix entries $1, 0, -1$ normalized by powers of $\sqrt{2}$ suffice to represent any quantum computation's input-output behavior. Interestingly enough, the first set is considered the easiest to engineer, with the second close behind, even though the last set has only real phases.

- Hadamard, $\boldsymbol{CNOT}$, and single-qubit $\boldsymbol{T}$-gates.
- Hadamard and $\boldsymbol{CS}$ gates.
- Hadamard and Toffoli gates.

7.8. Recalling problem 3.7 in chapter 3, define instead

$$\widetilde{\boldsymbol{U}} = \boldsymbol{R} \otimes \boldsymbol{I} + \boldsymbol{Q} \otimes \boldsymbol{R_x}(\pi).$$

Explain why $\widetilde{\boldsymbol{U}}$ is still unitary. Show that any measurement involving $\boldsymbol{U}$ can be simulated by measurement(s) involving $\widetilde{\boldsymbol{U}}$ instead.

7.9. Compute $\widetilde{\boldsymbol{S}}$ and show that it is the same as $\boldsymbol{CR_x}(\pi)$, that is, the controlled version of the product $\boldsymbol{XZ}$.

7.10. Now find $\widetilde{\boldsymbol{U}}$ with $\boldsymbol{U} = \boldsymbol{CS}$ instead.

7.11. Next show that the $\widetilde{\boldsymbol{U}}$ in problem 7.10 can be simulated by two Hadamard and two Toffoli gates.

7.12. For the drumroll, show that the $\sim$ operation commutes with matrix multiplication. You may find it easier first to argue the same thing for the mapping $\boldsymbol{U}'$ in problem 3.7, namely, that for any matrices $\boldsymbol{A}$ and $\boldsymbol{B}$,

$$(\boldsymbol{AB})' = \boldsymbol{A}'\boldsymbol{B}'.$$

7.13. Conclude that the observation in problem 7.8 about measurements in fact applies to entire quantum circuits of real-transformed gates. Hence, conclude that although Hadamard and Toffoli cannot directly simulate any gate with complex entries, they can simulate the measurements—and hence the outputs—of any circuit involving Hadamard and $\boldsymbol{CS}$ gates.

7.14. Conclude that, although the Quantum Fourier Transform cannot be approximated by Hadamard and Toffoli gates, because it has complex-number entries and the latter do not, the *measurement probabilities* of any (feasible) circuit that uses QFT can be closely approximated via *measurements* of a (feasible) circuit involving just Hadamard and Toffoli gates in place of QFT.

7.8 Summary and Notes

This chapter has given an overview of the structure of the next chapters and a peek at how quantum algorithms get their power: entanglement, forking, interference (i.e., cancellation), amplification (whose significance is raised by squaring), and the handle on algebra of size N that is given by primitives that scale with n. All of these elements have been hotly debated in more than a century of physics and philosophy. We have only briefly mentioned that entanglement was puzzling to Einstein, and we passed over "Schrödinger's Cat" without a sniff of paradox. We somewhat agree with Stephen Hawking's noted quip on the latter:

> When I hear of Schrödinger's cat, I reach for my gun.

What we really reach for is material on the engineering problem of building quantum computers that *scale*, which turns on the rate at which physical noise can "muss the hair" of our quantum animals. For anything as large as a cat, it may simply be too much, but if components can be made small enough, be isolated enough, and run fast enough, errors caused by "noise" may be minimal or correctable enough to enable real machines to work as the blueprints say. We hosted a year-long debate about this between mathematician Gil Kalai and computer physicist Aram Harrow on the *Gödel's Lost Letter* blog in 2012.

Still, we must concede that our own animal analogy for the workings and results of quantum algorithms has resorted to some weird features: duplicate Phils, Anti-Phils, mutant squaring. To express the intuition of David Deutsch for his first quantum algorithm, we would have to say further that each Hadamard fork creates not only a duplicate Phil but also a duplicate maze in its own parallel universe. Multiple Phils would somehow amplify and interfere with each other through the boundaries of these universes. We would respond more simply that it suffices to use the idea of summing squares, which goes back before Pythagoras, and that wave interference and the two-norm of vectors were studied even before the public birth of probability theory with Blaise Pascal. Still, we realize that such historical grounding does not prevent the debates from becoming deeper. Rather than itemize some of the myriad sources for these debates, we feel the best thing is to dive in and start covering Deutsch's algorithm.

A similar picturing of sum-over-paths to ours is in the survey by Aharonov (1998, 2008). Regarding the constant to divide by after "squaring Phils," we should say more precisely that it depends on the number of stages with *branching*; the Pauli $\boldsymbol{Z}$ matrix, for example, would give "cheese" but no branching. Matters like this are treated further in chapter 16 and its exercises. The simulation of Toffoli gates by two-qubit gates in the exercises is from Barenco et al. (1995). The quantum circuit diagram in problem 7.5 is the immediately first example in the tutorial for the `Qcircuit.tex` package, which we have used for this book. We took the diagram's code verbatim from the tutorial, except both it and the paper have a general double-controlled $\boldsymbol{U}$ in place of $\boldsymbol{CCX}$ for Toffoli, because the only property the construction needs is $\boldsymbol{V}^2 = \boldsymbol{U}$. Aharonov (2003), following on from Shi (2003), offers the basis for problem 7.8 and the exercises after it. These complete a proof of the quantum universality of Hadamard plus Toffoli gates, which formally justifies limiting attention to "Phil" and "Anti-Phil" in visualizing measurements and is the ground for theorems in chapter 16.

The reality of qubits is just one of many physical capabilities shown by experiments with entanglement. The one over sizable distances over Lake Geneva is by Salart et al. (2008) and is actually titled, "Testing Spooky Action at a Distance." Systems that use entanglement to verify that communications between banks have not been eavesdropped are already in practical use. Last and most, we should say that the "algorithm" emphasized by Feynman (1982, 1985) is the ability of quantum computers to simulate quantum processes in real time. We have not devoted a chapter to this algorithm, except insofar as this chapter serves.

8 Deutsch's Algorithm

Deutsch's algorithm operates on a Boolean function:

$$f\colon \{0,1\} \to \{0,1\}.$$

The goal is to tell whether the function is a constant by performing only **one** evaluation of the function. Clearly this is impossible in the classical model of computation, but the quantum model achieves this in a sense delineated below.

This problem is important for its historical significance because it was the first nontrivial quantum algorithm. It also shows that quantum algorithms can be more efficient than classical ones, even if the advantage in this case is minor. The classical solution requires two evaluations of the function, whereas the quantum solution requires only one. This is not an impressive difference, but it is there, and even this small difference suggested, correctly, that far greater differences would be possible. For this reason, Deutsch's algorithm retains its importance. So let's start to look at it in detail.

8.1 The Algorithm

We will present the algorithm as computing a series of vectors $\boldsymbol{a}_0, \boldsymbol{a}_1, \boldsymbol{a}_2, \boldsymbol{a}_3$, each of which is in the real Hilbert space $\mathbb{H}_1 \times \mathbb{H}_2$, where $\mathbb{H}_1$ and $\mathbb{H}_2$ are two-dimensional spaces. We index vectors in this space by xy, where x and y are single bits. For whichever Boolean function f is specified, recall again from section 4.3 that we work with its invertible extension, which we here symbolize as $f'(xy) = x(f(x) \oplus y)$. Thus, the "input" to the algorithm is really the choice of f as a parameter. The algorithm always uses the same input vector and goes as follows:

1. The initial vector is $\boldsymbol{a}_0$ so that $\boldsymbol{a}_0(01) = 1$.
2. The next vector $\boldsymbol{a}_1$ is the result of applying the Hadamard transform on each $\mathbb{H}_i$ of the space with $i = 1, 2$ separately.
3. Then the vector $\boldsymbol{a}_2$ is the result of applying $\boldsymbol{U}_{f'}$ where $f'(xy) = x(f(x) \oplus y)$.
4. The final vector $\boldsymbol{a}_3$ is the result of applying the Hadamard transform again, but this time only to $\mathbb{H}_1$.

Note that in case f is the identity function, f' becomes the Controlled-NOT function, and $\boldsymbol{U}_{f'}$ becomes the 4×4 ***CNOT*** matrix. Because f is the identity, we rename it $\boldsymbol{U}_I$. Similarly, we write $\boldsymbol{U}_X$, $\boldsymbol{U}_T$, and $\boldsymbol{U}_F$ for the cases f being the negation, always-true, and always-false function, respectively. Thus, the four possible matrices that incorporate the given function f are:

$$\boldsymbol{U_I} = \begin{bmatrix} 1 & 0 & 0 & 0 \\ 0 & 1 & 0 & 0 \\ 0 & 0 & 0 & 1 \\ 0 & 0 & 1 & 0 \end{bmatrix} \quad \boldsymbol{U_X} = \begin{bmatrix} 0 & 1 & 0 & 0 \\ 1 & 0 & 0 & 0 \\ 0 & 0 & 1 & 0 \\ 0 & 0 & 0 & 1 \end{bmatrix} \quad \boldsymbol{U_T} = \begin{bmatrix} 0 & 1 & 0 & 0 \\ 1 & 0 & 0 & 0 \\ 0 & 0 & 0 & 1 \\ 0 & 0 & 1 & 0 \end{bmatrix} \quad \boldsymbol{U_F} = \begin{bmatrix} 1 & 0 & 0 & 0 \\ 0 & 1 & 0 & 0 \\ 0 & 0 & 1 & 0 \\ 0 & 0 & 0 & 1 \end{bmatrix}$$

Note that the matrices $\boldsymbol{U_T}$ and $\boldsymbol{U_F}$ are unitary even though the always-true and always-false functions are not reversible. This illustrates the quantum trick of preserving the "x" argument of these functions as the first qubit and recording $f(x)$ in terms of its effect when exclusive-or'ed with the second qubit, y.

According to the algorithm, we will sandwich one of these four matrices between the $\boldsymbol{H}_2 \otimes \boldsymbol{H}_2$ matrix on the left and the matrix for $\boldsymbol{H}_2 \otimes \boldsymbol{I}$ on the right. The latter we saw in chapter 7, whereas the former is

$$\frac{1}{2}\begin{bmatrix} 1 & 1 & 1 & 1 \\ 1 & -1 & 1 & -1 \\ 1 & 1 & -1 & -1 \\ 1 & -1 & -1 & 1 \end{bmatrix}.$$

The chain of three matrices is applied to the start vector $\boldsymbol{a}_0 = \boldsymbol{e}_{01}$ on the right, producing in each of the four cases the vector $\boldsymbol{a}_3$. A measurement of $\boldsymbol{a}_3$ will then determine whether we are in one of the two constant cases, where $\boldsymbol{U_T}$ or $\boldsymbol{U_F}$ is used, or whether we have one of the other two cases $\boldsymbol{U_I}$ or $\boldsymbol{U_X}$, which represent the nonconstant functions f. The point again is that in contrast to classical algorithms, which need to call f twice to evaluate $f(0)$ and $f(1)$, the quantum algorithm can tell the difference with just one $\boldsymbol{U}_{f'}$ oracle matrix, where again f' is the "controlled" version of f.

8.2 The Analysis

First let's invite Phil—the mouse from chapter 7—to do the analysis. Given the input $\boldsymbol{e}_{01}$, he enters at 01. In figure 8.1, we see the same maze stage at left and right, which corresponds to a Hadamard gate on the first of two qubit lines. The stage for Hadamard on line 2 comes after it on the left. Next, one of the four matrices above is filled in the blank for $\boldsymbol{U}_{f'}$. Each is a permutation matrix, so its four corridors will run across with no branching, and we can already see what they do: U_F makes the corridors run straight across, U_X interchanges the top two, U_I interchanges the bottom two, and U_T swaps both.

Figure 8.1
Maze for Deutsch's algorithm.

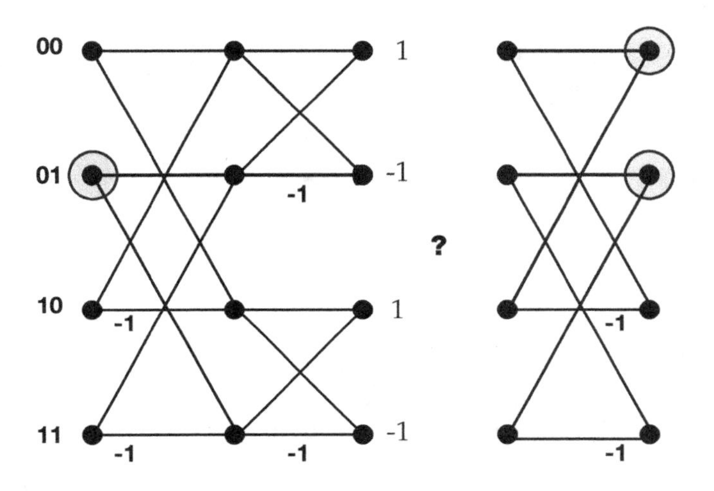

Phil starts running and splits into four by the time "he" reaches the gap: Phil at 00, Anti-Phil at 01, Phil at 10, and Anti-Phil at 11. Now we can visualize what happens when the missing stage is dropped in. Figure 8.2 shows the diagrams for the stages corresponding to the matrices U_I, U_X, U_T, U_F given above.

If the stage does U_F for the constant-false function, then the mice run straight across to the last stage. Then the two Phils can each run to exit at 00, so 00 has positive amplitude 2. The two Anti-Phils are likewise the only mice who can run to the exit for 01, and they amplify each other to give -2 there. Although this is negative, its square will still be 4, the same as for 00. Because there are three cheese stages, the divisor is $2^3 = 8$, so each outcome has probability 0.5. This already shows there cannot be any amplitude left over for outcomes 10 or 11, but let us verify from the maze: each gets a Phil and an Anti-Phil, which cancel.

Figure 8.2
Maze stages for possible queried functions.

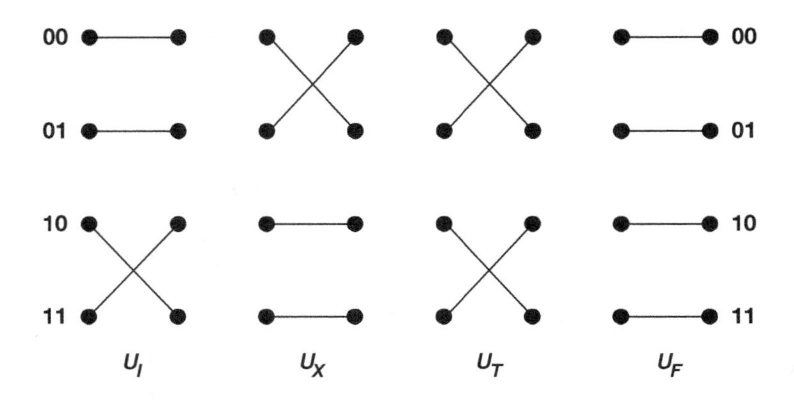

The matrix $\boldsymbol{U_T}$ swaps the top and bottom pairs of Phil and Anti-Phil, but this changes nothing in the above analysis, except now 00 has −2 and 01 has +2. So again a measurement can give only 00 or 01.

The other two matrices do only one of the swaps, however, and that changes the picture. Now Phil and Anti-Phil gang up to cancel the 00 and 01 outcomes, but they amplify for 10 and 11. Hence, a measurement certainly gives "1" in the first place. The set S corresponding to 0 gives rise to a 1-qubit measurement that perfectly distinguishes the constant-function and nonconstant cases.

To be rigorous—and because we said "analysis by Phil" does not *scale* as n grows—we need to do the linear algebra. We state our objective formally:

THEOREM 8.1 A measurement of the vector $\boldsymbol{a}_3$ will return $0y$, for some y, if and only if f is a constant function. Thus, Deutsch's algorithm tells whether f is constant using just one application of $\boldsymbol{U}_{f'}$.

Of course this theorem is the key: one application of f and one measurement will tell whether f is a constant function. To save multiplying out the 4×4 matrices for each of the four cases—a method that doesn't scale either—we use our notation indexing vectors $\boldsymbol{a}$ by $\boldsymbol{a}(00), \boldsymbol{a}(01), \boldsymbol{a}(10), \boldsymbol{a}(11)$. The proof depends on the following lemma, in which we use binary XOR on bits to denote a number.

LEMMA 8.2 The following are true:

1. For all xy, $\boldsymbol{a}_1(xy) = \frac{1}{2}(-1)^y$.
2. For all xy, $\boldsymbol{a}_2(xy) = \frac{1}{2}(-1)^{f(x)\oplus y}$.
3. For all xy, $|\boldsymbol{a}_3(xy)|^2 = \frac{1}{8}\left|(-1)^{f(0)} + (-1)^{f(1)\oplus x}\right|^2$.

Proof. Let us prove (1). It is clear that applying Hadamard gates independently yields

$$\boldsymbol{a}_1(xy) = \frac{1}{2}\sum_{t,u}(-1)^{x\cdot t}(-1)^{y\cdot u}\boldsymbol{a}_0(tu).$$

Thus, by the definition of $\boldsymbol{a}_0$,

$$\boldsymbol{a}_1(xy) = \frac{1}{2}(-1)^{x\cdot 0}(-1)^{y\cdot 1},$$

which is $\frac{1}{2}(-1)^y$.

Let us next prove (2). By definition of the matrix $\boldsymbol{U}_{f'}$ it follows that

$$\boldsymbol{a}_2(xy) = \boldsymbol{a}_1(x(f(x)\oplus y)) = \frac{1}{2}(-1)^{f(x)\oplus y}.$$

Let us finally prove (3). Again by definition of the Hadamard transform,

$$\begin{aligned}\boldsymbol{a}_3(xy) &= \frac{1}{\sqrt{2}}\sum_t(-1)^{x\cdot t}\boldsymbol{a}_2(ty),\\ &= \frac{1}{2\sqrt{2}}\sum_t(-1)^{x\cdot t}(-1)^{f(t)\oplus y}.\end{aligned}$$

Note that we can expand the sum and show that it is

$$\frac{1}{2\sqrt{2}}\left((-1)^{f(0)\oplus y} + (-1)^{x\oplus f(1)\oplus y}\right).$$

We can factor out the common term $(-1)^y$ to get the amplitude:

$$|\boldsymbol{a}_3(xy)|^2 = \frac{1}{8}\left|(-1)^{f(0)} + (-1)^{f(1)\oplus x}\right|^2.$$

□

Proof of Theorem 8.1. By lemma 8.2, $|\boldsymbol{a}_3(0y)|^2$ is

$$\frac{1}{8}\left|(-1)^{f(0)} + (-1)^{f(1)}\right|^2.$$

If f is constant, then this expression is equal to $\frac{1}{8}2^2 = \frac{1}{2}$. If f is not constant, then it is equal to 0. □

8.3 Superdense Coding and Teleportation

We digress from our algorithms involving unknown functions to show other actions with four alternatives. Both involve entanglement and begin with the Hadamard plus ***CNOT*** combination detailed in chapter 7, rather than with two Hadamard gates as in Deutsch's algorithm. They carry out the most basic forms of general constructions called **superdense coding** and **quantum teleportation**.

Both applications involve a physical interpretation and realization of *qubits*. As in the last chapter, let us talk of people named "Alice" and "Bob" across Lake Geneva from each other. First consider a general product state

$$\begin{aligned} \boldsymbol{c} &= (a_0\boldsymbol{e}_0 + a_1\boldsymbol{e}_1) \otimes (b_0\boldsymbol{e}_0 + b_1\boldsymbol{e}_1) \\ &= a_0b_0\boldsymbol{e}_{00} + a_0b_1\boldsymbol{e}_{01} + a_1b_0\boldsymbol{e}_{10} + a_1b_1\boldsymbol{e}_{11}, \end{aligned}$$

where $|a_0|^2 + |a_1|^2 = 1$ and $|b_0|^2 + |b_1|^2 = 1$. Here we can regard $\boldsymbol{a} = a_0\boldsymbol{e}_0 + a_1\boldsymbol{e}_1$ as a qubit wholly in the control of Alice and $\boldsymbol{b} = b_0\boldsymbol{e}_0 + b_1\boldsymbol{e}_1$ as a qubit owned by Bob, with $\boldsymbol{c} = \boldsymbol{a} \otimes \boldsymbol{b}$ standing for the joint state of the system. Thus far, so good.

The interpretation is that the identification and ownership of qubits applies even when the system is in a general pure state of the form

$$\boldsymbol{d} = d_{00}\boldsymbol{e}_{00} + d_{01}\boldsymbol{e}_{01} + d_{10}\boldsymbol{e}_{10} + d_{11}\boldsymbol{e}_{11}$$

with $|d_{00}|^2 + |d_{01}|^2 + |d_{10}|^2 + |d_{11}|^2 = 1$. In terms of quantum coordinates, Alice controls the first index, which plays d_{00}, d_{01} against d_{10}, d_{11}, while Bob controls the second index, which plays the even-index entries d_{00}, d_{10} in places 0 and 2 off against the odd entries d_{01}, d_{11} in places 1 and 3. There is one other partition that plays off two against two, the "outers" d_{00}, d_{11} versus the "inners" d_{01}, d_{10}. This playing-off can be achieved directly by a different kind of measurement that projects onto the transformed basis whose four elements are given by $\boldsymbol{e}_{00} \pm \boldsymbol{e}_{11}$ and $\boldsymbol{e}_{01} \pm \boldsymbol{e}_{10}$, each normalized by dividing by $\sqrt{2}$. This basis is named for John Bell, who proved a famous theorem showing that the statistical results of measuring entangled systems cannot be explained by deterministic theories with local interactions only.

The physical realization is that after converting our usual all-zero start state $\boldsymbol{e}_{00}$ to

$$\boldsymbol{d} = \frac{1}{\sqrt{2}}\boldsymbol{e}_{00} + \frac{1}{\sqrt{2}}\boldsymbol{e}_{11},$$

we really can give Alice a particle representing the first coordinate and shoot Bob across the lake an entangled particle representing the second coordinate. The experiments mentioned in the last chapter and below have demonstrated that Alice and Bob can for some time keep these particles in this joint state. Moreover, Alice is physically able to operate further on this state by matrix operators applied only to *her* qubit, that is, operators of the form $\boldsymbol{U} \otimes \boldsymbol{I}$ where $\boldsymbol{U}$ is a 2×2 unitary matrix.

In particular, let $\boldsymbol{U}$ be one of four things: (i) $\boldsymbol{I}$, (ii) $\boldsymbol{X}$, (iii) $\boldsymbol{Z}$, or (iv)

$$\boldsymbol{XZ} = \begin{bmatrix} 0 & -1 \\ 1 & 0 \end{bmatrix} = -i\boldsymbol{Y}.$$

Thus, Alice is applying one of the Pauli matrices discussed in the exercises to chapter 3. Let Alice do *one* of these four things and then shoot her qubit across the lake to Bob. Can Bob, now able to carry out multi-qubit operations such as **CNOT**, figure out which one she did? The answer is *yes*. What he does is "uncompute" the original entanglement and measure both qubits. Here is the whole system expressed as a quantum circuit, this time with a standard symbol for measurements at the end:

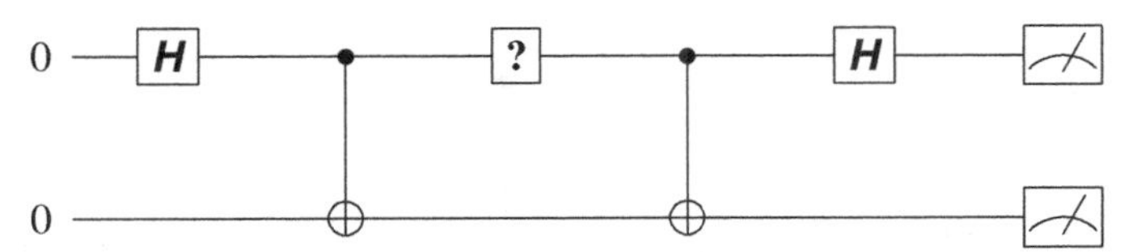

To show the similarity to the analysis of Deutsch's algorithm, we draw the corresponding maze diagram for the circuit with a missing stage, and the diagrams for the four possible stages Alice can insert, in figures 8.3 and 8.4.

In each case, two Phils congregate at one of the exit points, except in the $-i\boldsymbol{Y}$ case, when two Anti-Phils end at 11. Because the amplitude divisor is 2, this already entails that the Phils at the other exit points always cancel, but one may enjoy verifying this from the two figures. Hence, the measurement always gives the same exit point depending only on the operation Alice chose. The main point is that Alice's four choices lead to four different results, so that Bob is able to tell what Alice did.

Why might this be surprising? Bob has learned two bits of information as a result of the single qubit that Alice sent across the lake. This seems to say that the one qubit carried two classical bits of information. However, there was one previous connection between them—via the intermediary who gave them the entangled qubits to begin with. A result called **Holevo's theorem** expresses

Figure 8.3
Maze for superdense coding.

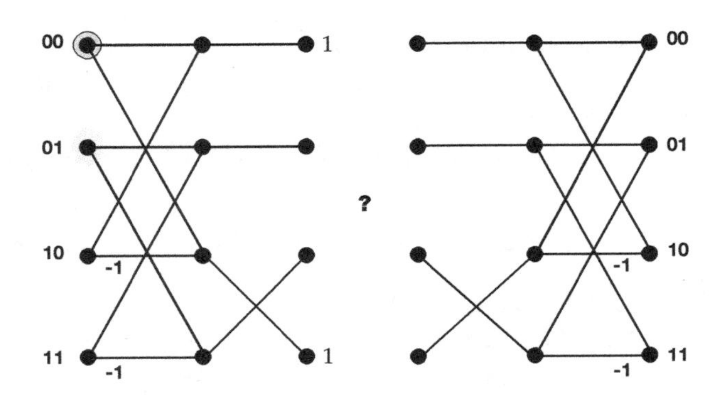

Figure 8.4
Maze stages for Pauli operators on qubit 1.

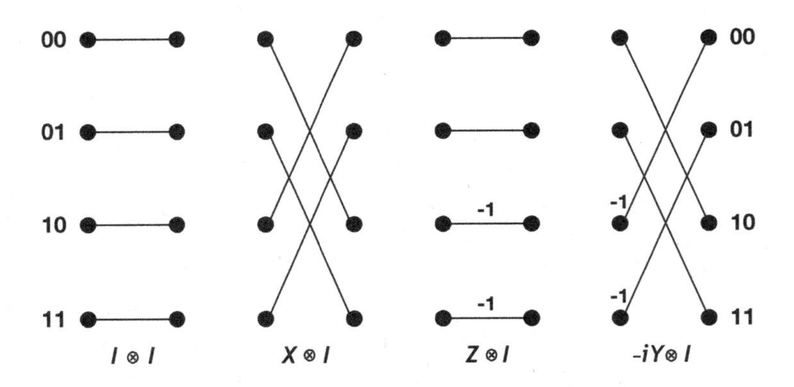

the deep principle that a total transmission of n qubits can carry no more than n bits of classical information. Thus, there must always have been some prior interaction between them or their environments to produce the entanglements. Once they are in place, however, Alice can electively transmit information at a classically impossible two-for-one rate—at the cost of consuming entanglement resources for each pair of bits. This explains the name *superdense coding*.

Quantum teleportation involves three qubits, two initially owned by Alice and one by Bob. Alice and Bob share entangled qubits as before, whereas Alice's other qubit is in an arbitrary (pure) state $\mathbf{c} = a\mathbf{e}_0 + b\mathbf{e}_1$. Alice has no knowledge of this state, and hence cannot tell Bob how to prepare it, yet entirely by means of operations on her side of the lake, she can ensure that Bob can possess a qubit in the identical state.

The following quantum circuit shows the operations, with $\mathbf{c}$ in the first quantum coordinate, Alice's entangled qubit second, and Bob's last. The circuit includes the Hadamard and **CNOT** gates used to entangle the latter two qubits.

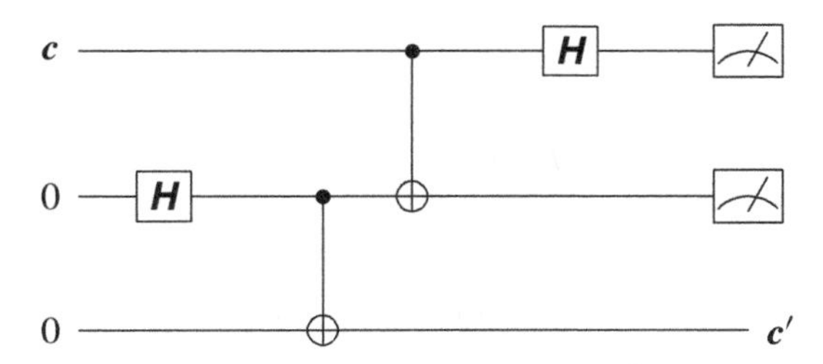

With this indexing, the start state is $\mathbf{c} \otimes \mathbf{e}_{00}$, which equals $a\mathbf{e}_{000} + b\mathbf{e}_{100}$. After the first two gates, the state is

$$\mathbf{c} \otimes \frac{1}{\sqrt{2}}(\mathbf{e}_{00} + \mathbf{e}_{11}),$$

with Alice still in possession of the first coordinate of the entangled basis vectors. The point is that the rest of the circuit involves operations by Alice alone, including the measurements, all done on her side of the lake. This is different from using a two-qubit swap-gate to switch the $\mathbf{c}$ part to Bob, which would cross the lake. No quantum interference is involved, so a maze diagram helps visualize the results even with "arbitrary-phase Phils" lined up at the entrances for $\mathbf{e}_{000}$ and $\mathbf{e}_{100}$ as shown in figure 8.5.

Because Bob's qubit is the rightmost index, the measurement of Alice's two qubits selects one of the four pairs of values divided off by the bars at the right. Each pair superposes to yield the value of Bob's qubit *after* the two measurements "collapse" Alice's part of the system. The final step is that Alice sends two *classical* bits across the lake to tell Bob what results she got, that is, which quadrant was selected by nature. The rest is in some sense the inverse of Alice's step in the superdense coding: Bob uses the two bits to select one of the Pauli operations $\mathbf{I}, \mathbf{X}, \mathbf{Z}, i\mathbf{Y}$, respectively, and applies it to his qubit $\mathbf{c}'$ to restore it to Alice's original value $\mathbf{c}$.

Figure 8.5
Maze for quantum teleportation.

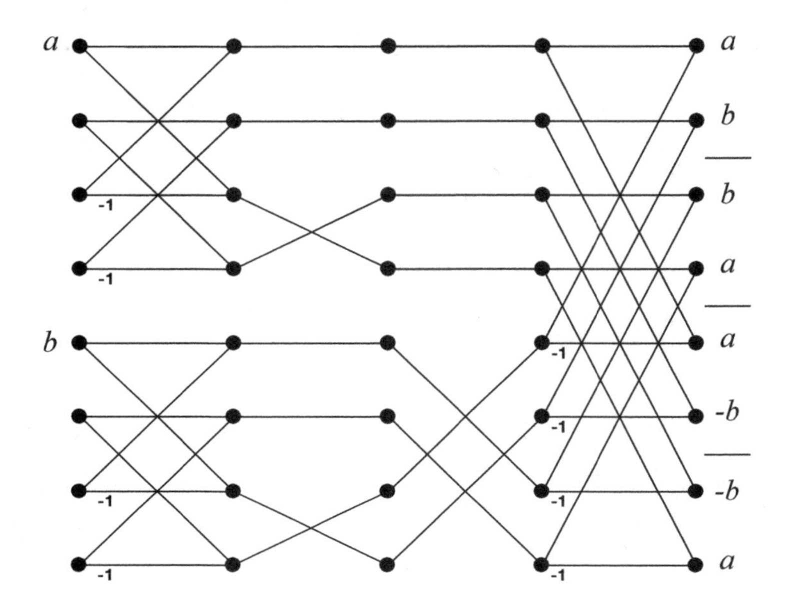

Neither Bob nor Alice is ever able to peek inside the qubit $\boldsymbol{c}$ to read the complex-number values of a and b or even get them right to more than a few uncertain bits of accuracy, amounting to at most one bit of solid information. This is already the essence of the natural law corresponding to Holevo's theorem. However, streams of qubits with prescribed values $\boldsymbol{c}$ can be generated, and experiments have shown that they can be received by Bob with high statistical fidelity over distances of many miles. This is still vastly far from the "Star Trek" dream of teleporting Alice across the lake, but already applications have sprung up to profit from this unexpected benefice of nature.

8.4 Problems

8.1. Prove that no classical algorithm can solve the problem in Deutsch's algorithm with one evaluation of f on Boolean arguments.

8.2. Directly show that the vector $\boldsymbol{a}_2$ in Deutsch's algorithm is a unit vector.

8.3. What happens if we apply the construction in Deutsch's algorithm to one of the other three basis vectors? What other conclusions can we draw?

8.4. Suppose Alice's quantum cellphone provider charges c cents per classical bit sent by text, q cents per qubit sent by "quantext," and e cents per entangled qubit. Work out the equations for superdense coding to be more cost-effective than just sending a classical text message. Then work out the conditions for teleportation to be cheaper (including the classical bits sent by text) than transmitting a qubit by quantext.

8.5. Can you devise a four-qubit circuit—or maze diagram with sixteen rows—in which Alice and Bob hold two qubits each, and at the end Bob has two copies of $\boldsymbol{c}$? Or can you do it with Alice and Bob having two entangled qubits, making three qubits for Alice and two for Bob overall, so that at the end Bob can do operations just on his two qubits to get his two copies of $\boldsymbol{c}$?

8.5 Summary and Notes

The original algorithm of Deutsch (1985) was not exactly the one presented. It solved the same problem, but it did not get it exactly right. Rather it got a probabilistic advantage even with one evaluation. The form now ascribed to him was established in more general form by Deutsch and Jozsa (1992), and we turn to that next.

Superdense coding originated with Bennett and Wiesner (1992), and quantum teleportation was discussed by Bennett, Brassard, Crépeau, Jozsa, Peres, and Wootters, (1993). Among many articles on experiments with teleportation, we mention two in *Nature*: Bouwmeester et al. (1997) and Marcikic et al. (2003). Holevo's theorem comes from Holevo (1973). One aspect is that whenever n-vertex (undirected, simple) graphs G are encoded with fewer than $\binom{n}{2}$ qubits, one per potential edge, the resulting quantum states $\boldsymbol{a}_G$ cannot always hold full information about G. Encodings $\boldsymbol{a}_G$ on (many) fewer qubits than edges can succeed only if the graphs G belong to families with regular structure or if the resultant smearing of information does not matter to approximation properties of the algorithm.

9 The Deutsch-Jozsa Algorithm

The Deutsch-Jozsa algorithm operates on a Boolean function:

$$f\colon \{0,1\}^n \to \{0,1\}.$$

The goal is to tell apart the cases where the function is *constant* or *balanced* by performing only **one** evaluation of the function. Here a function is *balanced* if it has the same number of 1's and 0's as output. If neither case holds then the output is immaterial. Clearly this goal is impossible in the classical model of computation, even with as many as 2^{n-1} evaluations of f on Boolean arguments. However, it is possible in the quantum model with just one evaluation, as we will see.

Deutsch's algorithm was important for being the first quantum algorithm, even though it only barely outperformed the classical one. The Deutsch-Jozsa algorithm shows that the improvement can be exponentially large. This is a huge advance over replacing two operations by one.

The claim of an exponential improvement is of a quantum algorithm compared with a deterministic classical algorithm. In the worst case, a classical algorithm might have to make an exponential number of evaluations of f before deciding whether it is balanced. However, a randomized algorithm could make this distinction in a constant number of evaluations provided we are happy to allow a small probability of making an error. Thus, this algorithm is another important step but still falls short of showing that quantum algorithms can be exponentially faster than classical algorithms if randomization is allowed for the latter. Still, the algorithm is important, and let's start to look at it in detail.

9.1 The Algorithm

We will present the algorithm as computing a series of vectors $\boldsymbol{a}_0, \boldsymbol{a}_1, \boldsymbol{a}_2, \boldsymbol{a}_3$, each which is in the real Hilbert space $\mathbb{H}_1 \times \mathbb{H}_2$, where $\mathbb{H}_1$ has dimension $N = 2^n$ and $\mathbb{H}_2$ has dimension 2.

We index vectors in this space by xy, where x is n bits and y is a single bit.

1. The initial vector is $\boldsymbol{a}_0$ so that $\boldsymbol{a}_0(0^n 1) = 1$. That is, $\boldsymbol{a}_0 = \boldsymbol{e}_{0^n 1}$.
2. The next vector $\boldsymbol{a}_1$ is the result of applying the Hadamard transform on each $\mathbb{H}_i$ of the space with $i = 1, 2$ separately.
3. Then the vector $\boldsymbol{a}_2$ is the result of applying $\boldsymbol{U}_{f'}$ where $f'(xy) = x(f(x) \oplus y)$.
4. The final vector $\boldsymbol{a}_3$ is the result of applying the Hadamard transform again, but this time only to $\mathbb{H}_1$.

9.2 The Analysis

We can try to analyze this with Phil the mouse starting from 0^n1, but this is where the failure of the mazes to *scale* starts to be felt. We can visualize that the initial Hadamard stages create a column that alternates Phil and Anti-Phil, much as in the analysis for Deutsch's original algorithm in the last chapter. We can further visualize that the permutation matrices for the two constant functions leave this arrangement the same or swap places for every Phil and Anti-Phil, and in both cases, the mice amplify the two outcomes that begin with 0^n and cancel the rest.

The *balanced* cases are harder to visualize, however, at least for us. How clear is it that they all cancel all the amplitude at 0^{n+1} and 0^n1? Here is where we hand the lead over to our linear-algebra indexing notation, for which we again state the goal as a theorem:

THEOREM 9.1 A measurement of the vector $\boldsymbol{a}_3$ will return $0^n y$, for some y, if and only if f is a constant function. Thus, the Deutsch-Jozsa algorithm distinguishes the cases of f being constant or balanced using only one evaluation of $\boldsymbol{U}_{f'}$.

Of course this theorem is the key: one measurement will work to tell whether f is a constant function. The proof depends on the following lemma. Note how it follows the logic of the previous chapter, where N was 2. Here, $N = 2^n$.

LEMMA 9.2 The following are true:

(1) For all x, y, $\boldsymbol{a}_1(xy) = \frac{1}{\sqrt{2N}}(-1)^y$.

(2) For all x, y, $\boldsymbol{a}_2(xy) = \frac{1}{\sqrt{2N}}(-1)^{f(x)\oplus y}$.

(3) For all x, y, $|\boldsymbol{a}_3(xy)|^2 = \frac{1}{2N^2}\left|\sum_t (-1)^{x\bullet t}(-1)^{f(t)}\right|^2$.

Proof. Let us prove (1). It is clear that applying the Hadamard gates independently yields

$$\boldsymbol{a}_1(xy) = \frac{1}{\sqrt{2N}}\sum_{t,u}(-1)^{x\bullet t}(-1)^{y\cdot u}\boldsymbol{a}_0(tu),$$

where we remind that $x \bullet t$ is the XOR-based inner product of Boolean strings, whereas $y \cdot u$ involves just single bits. Thus, by the definition of $\boldsymbol{a}_0$,

$$\boldsymbol{a}_1(xy) = \frac{1}{\sqrt{2N}}(-1)^{x\bullet 0}(-1)^{y\cdot 1},$$

which equals $\frac{1}{\sqrt{2N}}(-1)^y$.

Let us next prove (2). By definition of the matrix $\boldsymbol{U}_g$ it follows that

$$\boldsymbol{a}_2(xy) = \boldsymbol{a}_1(x(f(x) \oplus y)) = \frac{1}{\sqrt{2N}}(-1)^{f(x)\oplus y}.$$

Let us finally prove (3). Again by definition of the Hadamard transform,

$$\begin{aligned}\boldsymbol{a}_3(xy) &= \frac{1}{\sqrt{N}}\sum_t(-1)^{x\bullet t}\boldsymbol{a}_2(ty),\\ &= \frac{1}{\sqrt{2N}}\sum_t(-1)^{x\bullet t}(-1)^{f(t)\oplus y}.\end{aligned}$$

We can factor out the common term $(-1)^y$ to get the desired probability:

$$|\boldsymbol{a}_3(xy)|^2 = \frac{1}{2N^2}\left|\sum_t(-1)^{x\bullet t}(-1)^{f(t)}\right|^2.$$

□

Proof of Theorem 9.1. By lemma 9.2, $|\boldsymbol{a}_3(0^n y)|^2$ is

$$\frac{1}{2N^2}\left|\sum_t(-1)^{f(t)}\right|^2.$$

If f is constant, then this expression is equal to

$$\frac{1}{2N^2}\left|\sum_t(-1)^{0\bullet t}\right|^2 = \frac{1}{2}.$$

Thus, the two equivalent cases $y = 0, 1$ each have probability $\frac{1}{2}$, making it certain that the measurement yields $0^n y$. If f is not constant, then it is equal to

$$\frac{1}{2N^2}\left|\sum_t(-1)^{0\bullet t}(-1)^{f(t)}\right|^2 = 0.$$

The sum $\sum_t(-1)^{f(t)}$ is 0 because f is balanced, and so the measurement never yields $0^n y$. □

9.3 Problems

9.1. What is the classical worst-case complexity of this problem?

9.2. Why is $\sum_t (-1)^{f(t)}$ equal to 0 when f is balanced?

9.3. Suppose that f is *almost* balanced, i.e., that the number of outputs of 0 minus the number of outputs of 1 is $o(2^n)$. Can you still use the above algorithm?

9.4 Summary and Notes

This algorithm is of course due to Deutsch and Jozsa (1992). The clean demonstration of quantum capability for a task whose direct classical analogue is infeasible in the worst case as n grows drew attention to the power of quantum computation in the years preceding Shor's explosive discovery. The hunt for a task whose classical version can be posed as a *function* that lies outside classical randomized polynomial time led to Simon's algorithm, which however is likewise based on query access to a black-box function f. We gave a different angle on the Deutsch-Jozsa and Simon algorithms in two posts on the *Gödel's Lost Letter* blog titled "Quantum Chocolate Boxes" and "More Quantum Chocolate Boxes," and respectively posted at:

- http://rjlipton.wordpress.com/2011/10/26/quantum-chocolate-boxes/
- http://rjlipton.wordpress.com/2011/11/14/more-quantum-chocolate-boxes/

These posts raise the question of whether the classical analogue is limited in a way that is "unfair" for comparison to the algebraic resources enjoyed by the quantum algorithm.

10 Simon's Algorithm

Daniel Simon's algorithm detects a type of period in a Boolean function. Let

$$f\colon \{0,1\}^n \to \{0,1\}^n$$

be a Boolean function. We are promised that there is a "hidden vector" s such that for all y and z,

$$f(y) = f(z) \iff y = z \oplus s. \tag{10.1}$$

In case s is the all-zero vector, this means f is bijective, whereas for all other s, the promise forces f to be two-to-one in a particularly simple way. In the latter case, we say f is *periodic* with "period" s.

Simon's beautiful theorem is that s can be found by a polynomial-time quantum algorithm. In particular, the algorithm distinguishes the case f is 1-to-1 from the particular cases where f is 2-to-1. This is the real breakthrough, because it beats even randomized classical algorithms. Thus, this algorithm is important in showing the power of quantum algorithms. The problem is artificial and not itself important, but it showed the way. So let's look at it in detail.

10.1 The Algorithm

Simon's algorithm is different from the previous algorithms in that we need to run the quantum part many times to discover the value of s. Roughly, each "run" of the quantum routine gets more information about the value of s, and eventually we will be able to recover it. Happily, this recovery procedure is a classical algorithm and is quite simple: you just need to find a solution to a linear system.

Initially, we have no equations for s. As the algorithm runs, it will accumulate more and more equations. Eventually it will have enough equations to solve and "find" s.

The algorithm operates on vectors from $\mathbb{H}_N \times \mathbb{H}_N$, where $N = 2^n$. As before, we view each vector as indexed by xy, except now both x and y are n-bit Boolean strings. The main body repeats until a classically verifiable condition is met. The condition is that a set E of linear equations has a unique solution. The set E uses the dot-product $\bullet$ of Boolean strings with addition modulo 2.

1. Initialize E to the empty set.
2. While E does not have a unique solution, do:

2.1 Define the initial vector $\boldsymbol{a}$ by

$$\boldsymbol{a}(xy) = \begin{cases} \frac{1}{\sqrt{N}} & \text{when } y = f(x) \\ 0 & \text{otherwise.} \end{cases}$$

We showed how to build $\boldsymbol{a}$ feasibly given a box computing f in section 6.4.

2.2 The next vector $\boldsymbol{b}$ is the result of applying the Hadamard transform to the "x" part of $\boldsymbol{a}$,

2.3 Measure $\boldsymbol{b}$, which gives a concrete answer xy.

2.4 Add the equation $x \bullet s = 0$ to the set E of equations.

3. Solve the equations to obtain a unique s.
4. If $s = 0^n$ answer "f is bijective"; otherwise by the promise, we have found a nonzero s such that $f(y) = f(z)$ whenever $z = y \oplus s$, and we can output some such pair as witness to the answer "f is not bijective."

10.2 The Analysis

The analysis of the algorithm is based first on the observation that

$$\boldsymbol{b}(xy) = \frac{1}{\sqrt{N}} \sum_{t} (-1)^{x \bullet t} \boldsymbol{a}(ty).$$

This follows because it is the definition of applying the Hadamard transform to the first part of the space.

THEOREM 10.1 Given f and a hidden s satisfying the promise of equation (10.1), Simon's algorithm finds s in polynomial expected time.

The following main lemma is the key to understanding the algorithm.

LEMMA 10.2 Suppose that f is periodic with nonzero s. Then the measured x's are random Boolean strings in $\{0, 1\}^n$ such that $x \bullet s = 0$.

Proof. In this case, f is two-to-one. Define R to be the set of y such that there is an x with $f(x) = y$, i.e., R is the range of the function f. Note that R contains exactly one-half of the possible y values.

If y is not in R, then $\boldsymbol{b}(xy) = 0$, because no t makes $\boldsymbol{a}(ty)$ nonzero. If y is in R, then there are two values z_1 and z_2 so that $f(z_i) = y$ for each i. Further,

$$z_1 = z_2 \oplus s.$$

In this case,

$$\begin{aligned} \boldsymbol{b}(xy) &= 1/N((-1)^{x \bullet z_1} + (-1)^{x \bullet (z_1 \oplus s)}) \\ &= 1/N(-1)^{x \bullet z_1}\left(1 + (-1)^{x \bullet s}\right). \end{aligned}$$

This is either 0 or $\pm 2/N$ depending on whether or not $x \bullet s = 0$. Thus, in this case, we have:

$$\boldsymbol{b}(xy) = \begin{cases} \pm 2/N & \text{if } y \in R \text{ and } x \bullet s = 0; \\ 0 & \text{otherwise.} \end{cases}$$

The case where $\boldsymbol{b}(xy)$ is nonzero occurs exactly for half the x's and for half the y's. This is as it should be—otherwise, the norm of $\boldsymbol{b}$ would not be 1.

Finally, it follows that any measurement yields xy with x a random Boolean string so that $x \bullet s = 0$ as claimed. □

Proof of Theorem 10.1. By lemma 10.2, we accumulate random x so that $x \cdot s = 0$. Because a random vector avoids even an $(n-1)$-dimensional subspace with probability at least one-half, the expected number of trials to obtain a full-rank system is below $2n$, and the probability of eventual success is overwhelming. If we are in the $s = 0$ case, then we will quickly find that out as well. The last step, on solving for a nonzero s, is to generate and verify the witness for f not being 1-to-1. □

Note, incidentally, that the classical part of the algorithm gives $\{0,1\}^n$ a vector-space structure, with bitwise XOR serving as vector addition modulo 2. This contrasts with the quantum part of the algorithm using N-dimensional space for its own reckonings.

10.3 Problems

10.1. What is the classical complexity of this problem if one can only evaluate f as a black box? Even if we allow randomized algorithms?

10.2. Show how to construct the start vector $\boldsymbol{a}$ from the elementary vector $\boldsymbol{e}_0$.

10.3. Show that for any constant bit-vector s, $x \bullet s$ is a linear equation over the finite field $\mathbb{Z}_2$.

10.4. Show that $\boldsymbol{b}$ is a unit vector directly.

10.5. Show that with probability at least $1/2$, a set of m randomly selected Boolean n-vectors is likely to span the whole space, provided $m = \Omega(n \log_2 n)$.

10.6. Consider the integers modulo $m = 2^n$. For every $k \leq n$, the multiples of 2^k are closed under addition, and hence they form an additive subgroup (in fact, an *ideal*) $H = H^{(k)}$ of $\mathbb{Z}_m$. For any $j < 2^k$, the set $H_j = \{h + j \mid h \in H\}$ is a *coset* of H, with $H_0 = H$.

Now say that $f\colon \mathbb{Z}_m \to \mathbb{Z}_m$ obeys the *hidden-subgroup promise* if for some k, f is constant on every coset of $H^{(k)}$, and its 2^k values on the cosets H_j are all distinct. Design a quantum algorithm to compute the value of k in $n^{O(1)}$ time.

10.7. Give a classical algorithm for the task of problem 10.6. Can you bound its running time by a polynomial in n? Consider k in the neighborhood of $n/2$.

10.8. Note that f in problem 10.6 is periodic with period 2^k—that is, for all $x < m$, $f(x + 2^k) = f(x)$ (wrapping modulo m). Moreover, all values $f(h)$ for $x \leq h < x + 2^k$ are distinct. Suppose instead that this holds with a number r in place of 2^k, where r is not a power of 2. Does your quantum algorithm in problem 10.6 still work?

10.4 Summary and Notes

Simon's algorithm dates to 1992 and appeared in full in Simon (1997). It distinguishes whether a Boolean function is one-to-one from the case of its being a *special type of* two-to-one. Note that f can be two-to-one without having a period s. If there were a quantum algorithm that could distinguish one-to-one from two-to-one without any restrictions, then that would have consequences for the famous graph isomorphism problem. Namely, given two graphs, we can create functions that have hidden structure if and only if the graphs are isomorphic. A version of Simon's algorithm for this more general kind of periodicity would place the graph isomorphism problem into the class BQP, which we explore in chapter 16.

As hinted in problem 10.6, the idea of Simon's algorithm extends to a fairly wide range of so-called **hidden subgroup problems**. The situation is the same: we are given an oracle for an f that takes distinct constant values on each coset of a subgroup H of a given group G. The basic idea of Simon's algorithm works nicely when G is abelian and finds tough sledding when not. Indeed, if it works when G is the symmetric group, then graph isomorphism is solved in BQP.

11 Shor's Algorithm

The centerpiece of Peter Shor's algorithm detects a *period* in a function. Let

$$f\colon \mathbb{N} \rightarrow \{0, 1, \ldots, M-1\}$$

be a feasibly computable function. We are promised that there is a period r, meaning that for all x,

$$f(x+r) = f(x).$$

The goal is to detect the period, i.e., to determine the value of r. Actually, we need more than this promise. We also need that the repeating values

$$f(0), f(1), \ldots, f(r-1)$$

are all distinct. Some call this latter condition "injectivity" or "bijectivity." Possible relaxations of this condition are explored in the exercises, and overall its necessity and purpose are not fully understood.

Shor's beautiful result has many applications, including factoring integers. This is important because many researchers believe that factoring is *far* from being feasible for *classical* computation, and many popular crypto-systems and information-assurance applications base their security properties on this belief. In this chapter, we will discuss only the period-finding task; then in chapter 12, we will show how to use period-detection to factor integers.

11.1 Strategy

Shor's algorithm, like Simon's algorithm, has quantum and classical components that interlink. In its original form, which we present here, the quantum algorithm is a subroutine that is used to generate samples from an instance-specific distribution that seems hard to emulate classically. Theorem 16.2 will later allow making it into a "one-piece" quantum algorithm, but that is not how we think of it originally. Here is the overall strategy:

1. Given an n-bit integer M, which we suppose is a product of distinct primes, use classical randomness to generate an integer a between 1 and $M-1$. First do $\gcd(a, M)$ to allow for the tiny chance that a already shares a factor with M, in which case we're done. Otherwise form the function $f_a(x) = a^x \bmod M$, which then has a period r that we wish to compute. This r will divide the product of $p-1$ over all primes p dividing M and *may* help us find them. Many values of a and r are unhelpful, but with substantial probability, a will be chosen so that r is computed and yields a factor p.

2. Use the classical feasibility of modular exponentiation via repeated squaring (shown in problem 2.9) to prepare the functional superposition of $f_a(x)$ over all $x < M$.
3. Run the quantum part once and measure all qubits. The string formed by the first ℓ of them, where ℓ is about $2\log_2 M$, yields a particular integer x in binary encoding. With substantial probability, x is "good" as defined below.
4. Then classical computation is used to try to infer r from x. Either this succeeds and we go to the next step or it is recognized that x was not good. In the latter case, we go back to step 3, running the quantum routine again.
5. There is still a chance the value of r may be unsuitable—that is, that the original a was an unlucky choice. In this case, we must begin again in step 1. But otherwise the value of r provides the only needed input to a final classical stage that yields a verifiable solution to the problem about M.

There is a further important aspect of Shor's algorithm, whose details go beyond the scope of this primer. Like many other sources, we have stinted a little on details about the quantum Fourier transform being **feasible**. In its literal form, it involves angles in the complex plane that become exponentially small as n increases. We can create coarse approximations to these values using a few basic gates by a process hinted in the exercises of chapters 6 and 7. The ultimate game—not only in Shor's full paper but in our understanding of the "quantum power" that allows feasible solution to problems like factoring that may be classically hard—is how these approximations interplay with the classical techniques in stage 4. Note that stage 4 will also involve approximation. However, this concern was not part of Shor's original brilliant insight—it is rather the "engineering afterward" in which there is still room for more discoveries. If we agree *a priori* that the quantum Fourier transform in its pure form is feasible, then what follows becomes a complete proof that factoring is likewise feasible.

11.2 Good Numbers

Let Q be a power of two, $Q = 2^\ell$, such that $M^2 \leq Q < 2M^2$. Say an integer x in the range $0, 1, \ldots, Q-1$ is **good** provided there is an integer t relatively prime to the period r such that

$$tQ - xr = k, \quad \text{where} \quad -r/2 \leq k \leq r/2. \tag{11.1}$$

The whole idea of Shor is to break down the period detection into two parts, which are the two middle stages of the above strategy. That is, we use a quantum algorithm to generate values x that have an enhanced probability of being good. Then we use a classical algorithm to use the good x to find the period r. It helps first to know how many such x there are.

LEMMA 11.1 There are $\Omega(\frac{r}{\log\log r})$ good numbers.

Proof. The key insight is to think of (11.1) as an equation modulo r. Then it becomes

$$tQ \equiv k \bmod r,$$

where $-r/2 \leq k \leq r/2$. But as t varies from 0 to $r-1$, the value of k can be arranged to be always in this range, so the only constraint on t is that it must be relatively prime to r. The number of values t that are relatively prime to r defines Euler's *totient* function, which is denoted by $\phi(r)$. Note that for each value of t there is a different value of x, so counting t's is the same as counting x's. Thus, the lemma reduces to a lower bound on Euler's function. But it is known that

$$\phi(z) = \Omega(\frac{z}{\log\log z}).$$

Indeed, the constant in the Ω approaches $e^{-\gamma}$, where $\gamma = 0.5772156649\ldots$ is the famous Euler-Mascheroni constant. In any event, this proves the lemma. □

If r is close to M, then by choosing Q close to M rather than M^2, we would stand a good chance of finding a good x just by picking about $\log \ell$-many of them classically at random. However, this does not help when r is smaller. The genius of Shor's algorithm is that the quantum Fourier transform can be used to drive amplitude toward good numbers in all cases.

11.3 Quantum Part of the Algorithm

Shor's algorithm is like Simon's algorithm. This should not be too surprising because it was based on Simon's algorithm. The key difference is to use the Fourier transform $\mathbf{F}_Q$ in place of the Hadamard transform. Recall that we have $Q = 2^\ell$ and $M^2 \leq Q < 2M^2$. The reason that we have chosen Q so large is to ensure $Q/r > M$.

We view each vector as indexed by xy, where x and y are ℓ-bit Boolean strings. The above strategy controls this routine, which is like the inner part of Simon's algorithm:

1. The start vector $\boldsymbol{a}$ is the functional superposition of f, i.e.,

$$\boldsymbol{a}(xy) = \begin{cases} \frac{1}{\sqrt{Q}} & \text{when } y = f(x) \\ 0 & \text{otherwise.} \end{cases}$$

2. The next vector $\boldsymbol{b}$ is the result of applying $\mathbf{F}_Q$ to the "x" part of $\boldsymbol{a}$.
3. Measure $\boldsymbol{b}$, giving an answer xy from which we discard y.
4. Exit into a classical routine that tests whether x is a good integer—if so, continue with the classical stages given later or else repeat from step 1.

We remind readers that xr in (11.1) is ordinary numerical multiplication, whereas xy in vector indices is binary string concatenation. Although y is discarded, the injectivity condition ensures that for every good x, the superposition caused by $\mathbf{F}_Q$ will contribute exactly r-many y's. Together with lemma 11.1, this will give a little short of order-r^2 good *pairs* xy. Hence, it suffices to show that every good pair receives $\Omega(\frac{1}{r})$ of the amplitude, giving $\Omega(\frac{1}{r^2})$ in probability. The resulting $\Omega(\frac{1}{\log\log r})$ probability of getting a good number on each trial will be large enough for the classical part to expect to succeed after relatively few trials of the quantum part.

11.4 Analysis of the Quantum Part

The intuition is that the QFT creates power series out of many angles β. Each series creates a large locus of points $0, \beta, 2\beta, 3\beta, \ldots$ For most angles β, the locus spreads itself over the circle so that its average—which is obtained by summing the corresponding power series of complex numbers $\exp(ik\beta)$—is close to the origin. If β is close to an integer multiple of 2π radians, however, then the angles all stay close to 0 modulo 2π, and the average stays close to the complex number 1. These "good" β embody multiples of the unknown period r, and so the process will distinguish those x that yield such β. The way that r "pans out" like a nugget of gold is similar to what happens with s in Simon's algorithm.

Much of the analysis can be done exactly before we take estimates to bound the amplitudes. It suffices to show that the good cases collectively grab a non-trivial fraction of the probability—we do not need estimates when β is "bad" at all. Let us consider any pair xy where y is in the range of f. With $\omega = \exp(\frac{2\pi i}{Q})$, we have:

$$\boldsymbol{b}(xy) = \frac{1}{\sqrt{Q}} \sum_{u=0}^{Q-1} \omega^{xu} \boldsymbol{a}(uy) = \frac{1}{\sqrt{Q}} \sum_{u:f(u)=y} \omega^{xu} \boldsymbol{a}(uy) = \frac{1}{Q} \sum_{u \in f^{-1}(y)} \omega^{xu}.$$

The last $\frac{1}{Q}$ is not a typo—we have substituted the value of $\boldsymbol{a}(uy)$. Now take the first x_0 such that $f(x_0) = y$. Then by injectivity,

$$f^{-1}(y) = \{x_0, \; x_0 + r, \; x_0 + 2r, \; x_0 + 3r, \ldots\}.$$

The cardinality of this set up to $Q - 1$ is $T = 1 + \lfloor \frac{Q - x_0}{r} \rfloor$. This brings out the finite geometric series and enables us to apply the formula for its sum:

$$\begin{aligned} \boldsymbol{b}(xy) &= \frac{1}{Q} \sum_{k=0}^{T-1} \omega^{x(x_0 + rk)} \\ &= \frac{\omega^{xx_0}}{Q} \sum_{k=0}^{T-1} \omega^{xrk} \\ &= \omega^{xx_0} \frac{1}{Q} \left(\frac{\omega^{Txr} - 1}{\omega^{xr} - 1} \right). \end{aligned}$$

Note that when we take absolute values, the complex-phase factor ω^{xx_0} will go away because it is a unit. We can multiply by further such units to make the numerator and denominator have real values even before we take the norms, using the trick that $\exp(i\beta) - \exp(-i\beta) = 2\sin(\beta)$:

$$\begin{aligned} \boldsymbol{b}(xy) &= \omega^{xx_0 - xrT/2} \frac{1}{Q} \left(\frac{\omega^{Txr/2} - \omega^{-Txr/2}}{\omega^{xr} - 1} \right) \\ &= \omega^{xx_0 - xrT/2 + xr/2} \frac{1}{Q} \left(\frac{\omega^{Txr/2} - \omega^{-Txr/2}}{\omega^{xr/2} - \omega^{-xr/2}} \right) \\ &= \omega^{x(x_0 + (T-1)r)} \frac{1}{Q} \left(\frac{\sin(T \cdot \pi xr/Q)}{\sin(\pi xr/Q)} \right). \end{aligned}$$

Note that we canceled a factor of 2 both inside and outside the angles. This finally tells us that

$$|b(xy)|^2 = \frac{1}{Q^2} \frac{\sin^2(T \cdot \pi xr/Q)}{\sin^2(\pi xr/Q)}. \tag{11.2}$$

We will also use this equation in section 13.5. It looks strange that the right-hand side is independent of y, but recall that we did use the property that y is in the range of f—without needing that $y = f(x)$. By injectivity, we have r-many such y's for any particular x. To finish the analysis, we need to show:

1. When x is good, the right-hand side of (11.2) is relatively large.
2. The total probability on good *pairs* xy is $\Omega(\frac{1}{\log n})$, where n is the number of digits in M, which is high enough to give high probability of finding a good x in $O(\log n)$-many trials.
3. If x is good, then in classical polynomial time we can determine the value of r.

The second statement will follow quickly after the first, and we handle both in the next section.

11.5 Probability of a Good Number

We state a fact about sines that has its own interest. Note that 1.581 in radians is a little bit more than $\pi/2$ to leave some slack. We target the number 0.63247 because its square is just above 0.4.

LEMMA 11.2 For all $T > 0$ and angles $\alpha > 0$ such that $T\alpha \leq 1.581$,

$$\frac{\sin(T\alpha)}{\sin(\alpha)} > 0.63247T.$$

Proof. The well-known identity $\sin(\alpha) \leq \alpha$, which holds for all $\alpha \geq 0$, makes it suffice to show that

$$\frac{\sin(T\alpha)}{T\alpha} > 0.63247.$$

Consider the function $\frac{\sin(x)}{x}$ for $0 < x \leq 1.581$. Its derivative has numerator $x\cos(x) - \sin(x)$ and denominator x^2. For $\frac{\pi}{2} < x < 1.581$ the derivative is negative since $\cos(x)$ is negative. For $0 < x < \frac{\pi}{2}$ it is also negative because the inequality $x < \tan(x)$ holds there. Because its derivative is always negative in this range, the function is minimized at the upper boundary $x = 1.581$, where it has value

$$\frac{\sin(1.581)}{1.581} > \frac{0.9999479}{1.581} > 0.63247.$$

□

LEMMA 11.3 For all pairs xy with x good, assuming $M \geq 154$, the probability of the measurement step outputting x is bounded below by $\frac{0.4}{r^2}$.

Proof. Recall that x being good means there is an integer t such that $-\frac{r}{2} \leq tQ - xr \leq \frac{r}{2}$, and that we have $Q > Mr$ and $T = 1 + \lfloor \frac{Q-x_0}{r} \rfloor$, where $x_0 \leq r$. From above, using $|\sin(x)| = |\sin(-x)| = |\sin(x+\pi)|$, we have:

$$\begin{aligned} |\boldsymbol{b}(xy)|^2 &= \frac{1}{Q^2}\frac{\sin^2(T \cdot \pi xr/Q)}{\sin^2(\pi xr/Q)} \\ &= \frac{1}{Q^2}\frac{\sin^2(T \cdot \pi(\frac{xr}{Q} - t))}{\sin^2(\pi(\frac{xr}{Q} - t))} \\ &= \frac{1}{Q^2}\frac{\sin^2(T \cdot \pi\frac{xr-tQ}{Q})}{\sin^2(\pi\frac{xr-tQ}{Q})} \\ &= \frac{1}{Q^2}\frac{\sin^2(T \cdot \pi\frac{tQ-xr}{Q})}{\sin^2(\pi\frac{tQ-xr}{Q})}. \end{aligned}$$

Now by goodness, the angle $\alpha = \pi\frac{tQ-xr}{Q}$ is at most $\pi\frac{r}{2Q}$. Because $T \leq 1 + \frac{Q}{r}$, we have

$$T\alpha \leq \frac{\pi}{2} + \frac{\pi r}{2Q}.$$

Because we chose $Q > Mr$, we have $\frac{\pi r}{2Q} < \frac{\pi}{2M} < 1.581 - \frac{\pi}{2}$ using the condition $M \geq 154$. Thus, $T\alpha \leq 1.581$, so as to meet the hypothesis of lemma 11.2. This gives us what we needed to hit our round-number probability target:

$$|\boldsymbol{b}(xy)|^2 \geq \frac{1}{Q^2}(0.63247T)^2 > \frac{0.4}{r^2}.$$

□

COROLLARY 11.4 The probability of getting a good number on each trial of the quantum part is $\Omega(\frac{1}{\log\log M})$, indeed at least 1-in-$\log_2\log_2 M$.

Proof. For every good x, there are r-many different y's for which $f^{-1}(y)$ is a set of cardinality T in the analysis of section 11.4. Thus, by lemma 11.1, there are $\Omega(\frac{r^2}{\log\log r})$ *pairs xy* for which lemma 11.3 applies. A glance at the proof plus converting natural logs to logs base 2 makes the number of such pairs at least $\frac{2.53r^2}{\log_2 \log_2 r}$. Thus, the total probability of getting a good number is at least

$$\frac{2.53}{2.5 \log_2 \log_2 r} > \frac{1}{\log_2 \log_2 M} .$$

□

This finishes everything quantum in Shor's algorithm—the remaining sections of this chapter finish the entirely classical deduction of the exact period from a fairly small expected number of trials.

As an aside, it is worth noting that the QFT analysis can also be conducted in a "lighter" fashion without needing the geometric-series formula and phase trick giving a ratio of sines for lemma 11.2. We need only go back to the first summation formula for the probability:

$$|\boldsymbol{b}(xy)|^2 = \frac{1}{Q^2} \left| \sum_{k=0}^{T-1} \omega^{rk} \right|^2 = \frac{1}{Q^2} \left| \sum_{k=0}^{T-1} \exp\left(\frac{2\pi\, irk}{Q}\right) \right|^2 .$$

LEMMA 11.5 If θ is in the interval $[0, \pi]$, then $\sin(\theta) \geq 0$. Also if θ is in the smaller interval $[\pi/4, 3\pi/4]$, then $\sin(\theta) \geq \frac{\sqrt{2}}{2}$. □

The reason that bounds on the sin function are especially important is that one way to prove that a complex number $a + ib$ has a large absolute value is to show that its imaginary part b is large. Because this part of $\exp(i\theta)$ is $\sin(\theta)$, it follows that understanding sines will play a role in our bounds.

LEMMA 11.6 Suppose $0 \leq xr \bmod Q \leq r/2$ and $j \in \{0, 1, \ldots, \lfloor Q/r \rfloor - 1\}$. Then the imaginary part of

$$\beta = \exp\left(\frac{2\pi\, ixrj}{Q}\right)$$

is always non-negative, and for at least half of the values j, it is bounded below by a fixed constant.

Proof. We know that

$$0 \leq xr \bmod Q \leq \frac{r}{2} .$$

Let t and k be such that $0 \leq k \leq \frac{r}{2}$ and

$$0 \leq xr - tQ = k \leq \frac{r}{2} .$$

We need to give a lower bound for

$$\exp(2\pi i x r j/Q).$$

But $xr = k + tQ$, so this is equal to

$$\exp(2\pi i(k + tQ)j/Q),$$

which by periodicity of exp is equal to

$$\exp(2\pi i k j/Q).$$

The imaginary part of the exponential function is $\sin(2\pi kj/Q)$. It is always non-negative because $2kj \leq Q$, and so the angle lies in the interval $[0, \pi]$. This proves that β is non-negative.

Finally, provided j is bounded away from 0 and Q/r, it follows that each term contributes at least $c > 0$ for some absolute c. This proves the lemma. □

Thus, the entire sum is at least T multiplied by a constant c that is bounded away from 0, so

$$|\boldsymbol{b}(xy)|^2 \geq \frac{1}{Q^2} c^2 T^2 \geq \frac{c^2}{r^2},$$

whereupon the rest is similar to before.

11.6 Using a Good Number

Now we can finish the analysis of the inner classical routine in proving the third statement at the end of section 11.4.

LEMMA 11.7 If x is good, then in classical polynomial time, we can determine the value of r.

Proof. Recall that x being good means that there is a t relatively prime to r so that (by symmetry)

$$xr - tQ = k \qquad \text{where} \qquad -\frac{r}{2} \leq k \leq \frac{r}{2}.$$

Assume that $k \geq 0$; the argument is the same in the case it is negative. We can divide by rQ and get the equation

$$\left|\frac{x}{Q} - \frac{t}{r}\right| \leq \frac{1}{2Q}.$$

We next claim that r and t are unique. Suppose there is another t'/r'. Then

$$\left|\frac{t}{r} - \frac{t'}{r'}\right| \geq \frac{1}{rr'} \geq \frac{1}{M^2}.$$

But then both fractions are close, which makes Q smaller than M^2, a contradiction.

Because r is unique, it follows that t is too. So we can treat

$$xr - tQ = k$$

as an integer program in a fixed number of variables: the variables are r, t, and two slack variables used to state

$$-r/2 \leq k \leq r/2$$

as two equations. While integer programs are hard in general, for a fixed number of variables, they are solvable in polynomial time. This proves the lemma. □

Usually this lemma is proved without recourse to integer programming. Instead, most sources use the special structure of the equation and argue that it can be solved by an elementary result in number theory. This comes from a classical problem in Diophantine approximation theory: given a fraction x/Q, find the best approximation to it with the denominator of a certain size. This is exactly what is needed here, and it can be done by technology based on continued fractions.

11.7 Continued Fractions

This section is optional. If you believe that an integer program in a fixed number of variables is easy to solve, then there is no need to read on about this alternative method. Or if you know about continued fractions, then this is a repeat for you. However, if you wish to see how the approximation problem can be solved, then read away.

Here is a classic continued fraction:

$$a + \cfrac{1}{b + \cfrac{1}{c + \cfrac{1}{d + \ddots}}}$$

Given any real number α, the main result of the theory of continued fractions is that a series of fractions can be generated that get closer and closer to α. Indeed, if α is a rational number, then the sequence always terminates. Let $\frac{a_k}{b_k}$ be the k-th such fraction. The process is called the **continued fraction algorithm**, and its analysis is conveyed by the following theorem statement.

THEOREM 11.8 The fraction $\frac{a_k}{b_k}$ can be generated in at most a polynomial number of basic arithmetic operations. Further, the distance from α to this fraction decreases exponentially fast. Also it is the best approximation to α in the following sense: if

$$|\alpha - \frac{c}{d}| \leq |\alpha - \frac{a_k}{b_k}|,$$

then $d \geq b_k$. □

Let's see whether this is enough to solve the approximation problem we face. Let $\alpha = \frac{x}{Q}$. Suppose we run the continued fraction algorithm until $\frac{a}{b}$ is within $\frac{1}{2Q}$ of α. We argue that b must equal r. Suppose that $b < r$. We know that

$$|\frac{t}{r} - \frac{a}{b}| \leq \frac{1}{Q}$$

because both terms are close to α. Then it follows that

$$|ar - bt| \leq \frac{br}{Q} < 1.$$

This implies that $b = r$. Thus, by continued fractions, one can compute the period exactly with effort bounded by a polynomial in the number of bits in M.

11.8 Problems

11.1. Suppose you have a routine R that correctly computes the period r of any given function $f\colon [N] \to [N]$ only when r is *odd* and works in $O(\log N)^2$ time. Create a routine R' that works for even r as well. What is its running time?

11.2. Now suppose that R outputs the correct r with probability (at least) $3/4$, outputting "fail" otherwise. What running time must your R' have now?

11.3. Show that if period detection is feasible for functions f that violate injectivity, then SAT can be solved by a polynomial time quantum algorithm.

11.4. Suppose that f violates the injectivity assumption only by having one value appear twice. Does Shor's algorithm still work?

11.5. In the alternate analysis at the end of section 11.5, what value of the constant c can you achieve? Compare it with the constant 0.4 obtained in lemma 11.3.

11.6. For $M = 21$, how many values a relatively prime to M are there? Chart the periods r that you get for each one.

11.7. This problem is for those who know continued fractions or wish to learn more. Show that the continued fractions generated in the theorem need not be the best approximation to the number.

11.9 Summary and Notes

Shor's brilliant result appeared in the 1994 FOCS conference (Shor, 1994) and then in full journal form (Shor, 1997). It did more than almost anything else to create the excitement around the field of quantum algorithms. The ability to use it to break cryptographic systems captured the imagination of many researchers and the concern of many others.

The algorithm has been generalized in modest ways. The first of us jointly showed (Boneh and Lipton, 1996) that the injectivity assumption could be relaxed to allow any value to occur a polynomial in n times. Later even stronger generalizations were found for period detection. The theorem that integer programs with a fixed number of variables are polynomial-time solvable was proved by Lenstra (1983).

The full article (Shor, 1997) cites work (Coppersmith, 1994, Griffiths and Niu, 1996) showing that the algorithm works even when the quantum Fourier transform is replaced by a fairly coarse approximation. Note that the QFT ostensibly requires a principal N-th root ω_N of unity for exponential-size N. No device is fine enough to carry out a rotation by ω_N for large N, but these and other sources have shown that a series of larger steps, each of moderate precision, can achieve the same mathematical effect we proved using the exact QFT. There are, however, still extensive debates about the feasibility of all this in practice, which we have referenced in numerous posts on the *Gödel's Lost Letter* blog. The focus on $M = 21$ in the exercises, recalling section 1.4, comes because at this time of writing, 21 is the highest number for which a practical run of Shor's algorithm has been claimed, although even this has been questioned (Smolin et al., 2013).

12 Factoring Integers

In this chapter, we will present the most famous application of the period finding algorithm of Shor: the ability to factor integers in quantum polynomial time. The reduction of factoring to period discovery is really a nice example of computational number theory, one that was known a decade before Shor applied it. His genius was in the realization that he could compute periods fast via quantum algorithms.

12.1 Some Basic Number Theory

We need some standard facts and notation from elementary number theory. As usual $x \bmod M$ is the residue of x modulo M, and $x \equiv y \bmod M$ means that x and y have the same residue modulo M. The greatest common divisor of x and y, written $\gcd(x, y)$, is the largest natural number that divides both x and y, and can be found via Euclid's algorithm in time quadratic in the lengths of x and y. The numbers $\{x \mid \gcd(x, M) = 1\}$ form a group under multiplication modulo M. If $\gcd(x, y) = 1$, then we say that x and y are relatively prime to each other. Every element x relatively prime to M has a finite smallest number ℓ so that $x^\ell \equiv 1 \bmod M$. We will use $\mathrm{ord}_M(x)$ to denote this number.

If p is prime, then the nonzero numbers modulo p, that is, the numbers $1, \ldots, p-1$, form a *cyclic* group under multiplication. Being cyclic means that they can all be written as powers of some element. One further important fact is the so-called *Chinese remainder theorem*: given distinct primes p, q and any elements x, y,

$$x \equiv y \bmod pq \iff x \equiv y \bmod p \quad \text{and} \quad x \equiv y \bmod q.$$

We also need to define quadratic residues and state Euler's criterion. It suffices to define them modulo an odd prime p. A number a, $1 \leq a \leq p-1$, is a **quadratic residue** (mod p) if there is an integer x such that x^2 is congruent to a modulo p. **Euler's criterion** states that this is true if and only if

$$a^{\frac{p-1}{2}} \equiv 1 \pmod{p}.$$

For quadratic nonresidues, the right-hand side is -1 modulo p. It is important again that by repeated squaring, a classical algorithm can compute the left-hand side in time polynomial in $\log p$ (see problem 2.9).

Another way to view what is going on is that the quadratic residues are the even powers of any generator of the cyclic group. Note that it is possible to decide whether a is an even power without needing to compute a generator g

or take logarithms to base g modulo p. The latter task, called the *discrete logarithm problem* (see problem 12.5 below), is also solved by Shor's algorithm, but not even quantum methods are known to be able to *find* a generator with high probability in polynomial time.

12.2 Periods Give the Order

Let's turn now to the question of factoring the number $M = pq$ where p and q are distinct odd primes. The general case works essentially in the same way, so it is reasonable to prove only this special case. Also this is the case of most interest to cryptography, so even if we could only do this case that would be important, but the methods easily can handle the case when M is divisible by many primes.

Define the following function $f_a(x) = (a^x \bmod M)$. This function has several key properties. It is periodic because

$$f(x+r') = f(x)$$

for any x, where $r' = (p-1)(q-1)$. This follows from the fact that $a^{p-1} \equiv 1 \bmod p$ and $a^{q-1} \equiv 1 \bmod q$ and from the Chinese remainder theorem. Thus, the function must have a minimal period r that divides r'.

The r—not the r'—is the value that Shor's algorithm returns. Even to get r, we must prove that the function's values $f_a(0),\dots,f_a(r-1)$ are distinct. By way of contradiction, suppose that

$$f_a(x) = f_a(y),$$

where $0 \leq x < y \leq r-1$. By definition it follows that

$$(a^x \bmod M) = (a^y \bmod M),$$

so that $a^x \equiv a^y \bmod M$. Thus, $a^{y-x} = 1 \bmod M$, and it follows that there is a smaller period than r, which is a contradiction.

12.3 Factoring

Suppose that $M = pq$ where p, q are odd primes; the general case is similar. There really is one theme that is used over and over in most methods of factoring: try to construct an integer x so that p divides x and q does not. Then the value of $\gcd(x,M)$ will equal p, and M is factored. Of course the roles of p and q can be exchanged—the key is that one divides x and the other does not. We exploit the ability to compute the gcd of two integers in polynomial time.

Let's look at how M can be factored provided we come upon a multiple r of $p-1$ or $q-1$. Define $A(r,p,q)$ to be true when only **one** of $p-1$ and $q-1$ divides r. Also define $B(r,p,q)$ to be true when **both** divide r, and furthermore

$$\frac{r}{(p-1)} \quad \text{and} \quad \frac{r}{(q-1)}$$

are both odd numbers. The rationale for these definitions is the following two lemmas:

LEMMA 12.1 There is a randomized algorithm $A^*(r,M)$ that factors M with probability at least one-half, provided $A(r,p,q)$ is true.

Proof. Assume that $A(r,p,q)$ is true and $p-1$ divides r. Pick a random a in $\{1,\ldots,M-1\}$ and compute $\gcd(a^r-1,M)$. We claim that at least half the time this is a factor of M. Picking a by the Chinese remainder theorem is equivalent to picking a modulo each prime separately. Now $a^r \equiv 1 \bmod p$ because $a^{p-1} \equiv 1 \bmod p$ and $p-1$ divides r. So we need to show that

$$a^r \equiv 1 \bmod q$$

is true only at most half the time. Because $q-1$ does not divide r, there is some b so that $b^r \not\equiv 1 \bmod q$, using that $\mathbb{Z}_q^*$ is a cyclic group of order $q-1$. But then the set of b so that $b^r \equiv 1 \bmod q$ is a proper subgroup, and so has at most half the elements modulo q. This proves the lemma. □

LEMMA 12.2 There is a randomized algorithm $B^*(r,M)$ that factors M with probability at least one-half, provided $B(r,p,q)$ is true.

Proof. Assume that $B(r,p,q)$ is true. Pick a random a in $\{1,\ldots,M-1\}$ and compute $\gcd(a^{r/2}-1,M)$. We claim that at least half the time this is a factor of M. Because $B(r,p,q)$ is true, there is an ℓ so that $(p-1)\ell = r$ and ℓ is odd and an m so that $(q-1)m = r$ and m is also odd. Among values a besides multiples of p or q, which are filtered out by the initial $\gcd(a,M)$ step, a acts as a random draw from $\{1,\ldots,p-1\}$ and $\{1,\ldots,q-1\}$ simultaneously. Hence at least half the time it will be a quadratic residue modulo one of the primes and a nonresidue modulo the other—suppose p and q respectively. Then,

$$\begin{aligned} a^{(p-1)/2} &\equiv 1 \bmod p \\ a^{(q-1)/2} &\equiv -1 \bmod q \end{aligned}$$

by Euler's criterion. The last step is to note that by definition of ℓ and m this is the same as:

$$\begin{aligned} a^{(p-1)\ell/2} &\equiv 1 \bmod p \\ a^{(q-1)m/2} &\equiv -1 \bmod q. \end{aligned}$$

Hence, we can apply the same reasoning as for lemma 12.1 to obtain a factor at least half the time. □

Thus, our goal is to find an r so that $A(r,p,q)$ or $B(r,p,q)$ is true. We start with r so that $p-1$ divides r. Define $r_i = r/2^i$, and let k be so that r_k is odd. We run $A^*(r_i, M)$ for all $i = 0, \ldots, k$. Then we run $B^*(r_{k-1}, M)$. If any try yields a factor, we are done. Otherwise, we try the process again. The final point is that this process works at least one-half the time.

Let's prove that. Initially, $p-1$ divides r by assumption. Thus, $q-1$ must also or $A(r,p,q)$ is true. Assume that $p-1$ and $q-1$ both divide r_i for all $i = 0, \ldots, k-1$, with k as above, and let at least one fail to divide r_k. If only one fails, then $A(r_k,p,q)$ is true, and we are done. So they both must fail to divide r_k. Note that $(p-1)\ell = r_{k-1}$ for some ℓ. The value of ℓ must be odd because $p-1$ fails to divide r_k. In the same way, it follows that $(q-1)m = r_{k-1}$ for some odd m. So it follows that $B(r,p,q)$ is true. This will prove that we have at least a one-half chance to find a factor—which for $M = pq$ will be p or q itself.

Because it is simple to verify that the number we get divides M, we can do $O(\log n) = O(\log\log M)$ trials of the entire algorithm to ensure success probability at least $3/4$. Further trials can *amplify* the success probability close to 1, technically pushing the theoretical failure probability below 2^{-n^c} for any preset exponent c. Thus, everything in this and the previous chapter combines to prove:

THEOREM 12.3 Given any integer M, Shor's algorithm finds a factor of M with high probability in quantum polynomial time. □

12.4 Problems

12.1. Show that for any odd prime p, if x and y are both quadratic *non*residues modulo p, then xy *is* a quadratic residue modulo p.

12.2. Suppose that $a^k \equiv 1 \bmod p$ and $b^\ell \equiv 1 \bmod p$ for some odd prime p and some k, ℓ. Can k and ℓ be relatively prime?

12.3. Show what the classical part of Shor's algorithm does on $M = 15$ given $r = 8$.

12.4. Show what the classical part does on $M = 21$ with $r = 12$.

12.5. Let g be a generator of the cyclic group $\mathbb{Z}_p^*$ of the integers $1, \dots, p-1$ under multiplication modulo p. Define the **discrete logarithm** (base g) of any $x < p$ to be the unique number $r < p$ such that $g^r = x$ in $\mathbb{Z}_p^*$. Now given p, g, x with the task of finding the unknown r, define $f_x(a, b) = g^a x^{-b}$. Show a sense in which f is periodic with period r, and find a relevant Abelian "hidden subgroup" of $\mathbb{Z}_p^* \times \mathbb{Z}_p^*$. (If you are ambitious, go on to reprise the strategy of chapter 11 to compute r in quantum expected polynomial time, thus solving the discrete logarithm problem.)

12.5 Summary and Notes

That getting the order is enough to factor traces back at least to 1984 (Bach et al., 1984). Our exposition of its theorem that given any multiple of $p - 1$ it is possible to factor M also appeared in a post by us on the *Gödel's Lost Letter* blog: "A Lemma on Factoring," http://rjlipton.wordpress.com/2011/12/10/a-lemma-on-factoring/.

13 Grover's Algorithm

The problem that Grover's algorithm solves is finding a "needle in a haystack." Suppose that we have a large space of size N, and one of the elements is special. It may be a solution to some problem that we wish to solve—one we could verify if we knew it. Then a classical algorithm in worst case would have to examine all the elements, whereas even a randomized algorithm expects to look at $N/2$ elements.

The power of quantum algorithms, as discovered by Lov Grover, is that $N/2$ can be improved to $O(N^{1/2})$. Compared with a classical random-search algorithm, the factor $1/2$ goes into the exponent, which is a huge improvement. We will now explain how and why the algorithm works.

13.1 Two Vectors

Define the "hit vector" $\boldsymbol{h}$ by $\boldsymbol{h}(x) = 1$ if x is the solution, or if x belongs to a possibly-larger set S of solutions, and $\boldsymbol{h}(x) = 0$ otherwise. Provided the number k of solutions is nonzero, dividing by $\sqrt{k}$ makes $\boldsymbol{h}$ a unit vector, and hence a legal quantum state. If we could build this state, then measurement would yield a solution with certainty.

The goal is to build a state $\boldsymbol{h}'$ close enough to $\boldsymbol{h}$ so that measuring $\boldsymbol{h}'$ will yield a solution with reasonable probability. The issue is that if we prepare a state $\boldsymbol{a}$ at random, then it is overwhelmingly likely to be far from $\boldsymbol{h}$. Measuring a random $\boldsymbol{a}$ is like guessing a $y \in [N]$ at random, and the probability k/N of its success is tiny unless k is huge.

What Grover's algorithm does is start with a particular vector $\boldsymbol{j}$ and jiggle it in a way that "attracts" it to $\boldsymbol{h}$. How can we do this if we don't know anything about $\boldsymbol{h}$ in advance? Actually, we do know something: *all entries of* $\boldsymbol{h}$ *that are indexed by solutions have the same value*. Moreover, and arguably more important, the entries corresponding to *non*solutions also agree on their value. Call these two statements together the *solution-smoothness* property. Now the $\boldsymbol{j}$ we start with has all of its entries equal to $1/\sqrt{N}$, which guarantees solution-smoothness even though we have no idea where the solutions (and nonsolutions) are. By linearity, it follows that:

> Every vector $\boldsymbol{a}$ in the two-dimensional subspace spanned by $\boldsymbol{h}$ and $\boldsymbol{j}$ has the solution-smoothness property.

Having those equal entries, as we saw in section 5.5, enables the *reflection* of any such $\boldsymbol{a}$ about $\boldsymbol{h}$ to be computed via the Grover oracle. Actually, recall that if S is the set of which $\boldsymbol{h}$ is the characteristic vector, then reflection about $\boldsymbol{h}$ needs the Grover oracle of the *complement* of S. However, we can instead complement $\boldsymbol{h}$ in the following sense: Define the "miss vector" $\boldsymbol{m}$ to have entry $1/\sqrt{N-k}$ for each nonsolution, 0 for each solution. Then $\boldsymbol{m}$ also belongs to the subspace, because

$$\boldsymbol{m} = \frac{1}{\sqrt{N-k}}(\sqrt{N}\boldsymbol{j} - \sqrt{k}\boldsymbol{h}),$$

and importantly, it is orthogonal to $\boldsymbol{h}$. Now reflection about $\boldsymbol{m}$ uses the original Grover oracle $\boldsymbol{U}_S$, which we recall is defined by

$$\boldsymbol{U}_S[x,x] = \begin{cases} -1 & \text{if } x \in S; \\ 1 & \text{otherwise,} \end{cases}$$

with all off-diagonal entries zero. From theorem 5.6 in section 5.4, which was proved in section 6.5, we know that computing the Grover oracle is feasible. This is because *testing* whether a given x is in S is carried out by a feasible Boolean function $f(x)$, notwithstanding our difficulty of *finding* any $x \in S$. This is how a quantum algorithm is able to avail itself of information about $\boldsymbol{h}_S$, information that is not as trivial as we might have supposed.

We also know that reflection about $\boldsymbol{j}$ is a feasible unitary operation. These two reflection operations supply the "jiggle" that we need. Finally, what helps the analysis is that $\boldsymbol{h}$ and $\boldsymbol{m}$ form an orthogonal basis for $\mathbb{H}$, and it will be convenient to describe the action geometrically with respect to this basis.

This is our first foray outside the standard basis of any vector space, but having only two dimensions makes it easy. Think of $\boldsymbol{m}$ as the "x-axis" and $\boldsymbol{h}$ as the "y-axis," and note that $\boldsymbol{j}$ is somewhere in the positive region between $\boldsymbol{m}$ and $\boldsymbol{h}$. That is, $\boldsymbol{j}$ makes an angle α with $\boldsymbol{m}$ such that $0 < \alpha < \frac{\pi}{2}$. The cosine of α is given by the inner product of $\boldsymbol{j}$ with $\boldsymbol{m}$, which depends only on k:

$$\langle \boldsymbol{j}, \boldsymbol{m} \rangle = \sqrt{\frac{N-k}{N}}.$$

Put another way, $\sin^2(\alpha) = \frac{k}{N}$, which was just the success probability of random guessing. But after initializing $\boldsymbol{a}$ to $\boldsymbol{j}$, if we can rotate $\boldsymbol{a}$ to make its angle θ with $\boldsymbol{m}$ to be close to $\frac{\pi}{2}$, then $\sin^2(\theta)$, which is always the success probability that measuring $\boldsymbol{a}$ gives a valid solution, will be close to 1. The algorithm achieves this by the geometrical principle that *reflections around two different vectors yield a rotation.*

13.2 The Algorithm

We first state the algorithm supposing that the number k of solutions is known. Indeed, Grover originally presented his algorithm in the case where k is foreknown to be 1, that is, when there is a unique solution. Once we understand the mechanism, we will see what to do when k is not known in advance.

1. Initialize the vector $\boldsymbol{a}$ to be the start vector $\boldsymbol{j}$.
2. Compute $\alpha = \sin^{-1}(\sqrt{\frac{k}{N}})$ and $t_k = \lfloor \frac{\pi}{4\alpha} \rfloor$.
3. Repeat the following **Grover iteration** t_k times:
   3.1 Apply $\boldsymbol{Ref_m}$ to $\boldsymbol{a}$ via the Grover oracle $\boldsymbol{U}_S$, obtaining the vector $\boldsymbol{a}'$.
   3.2 Apply $\boldsymbol{Ref_j}$ to $\boldsymbol{a}'$, obtaining the new value of $\boldsymbol{a}$.
4. Measure the final state $\boldsymbol{a}$, giving a string $x \in \{0,1\}^n$.
5. If $x \in S$ stop—we have found a solution. Otherwise repeat the entire process.

Tacit here is that if $t_k < 1$, then the inner loop falls through and we measure right away. This happens only when $\frac{\pi}{4\alpha}$ is just below 1, i.e., $\alpha \geq \frac{\pi}{4}$, so $N/k \geq \frac{1}{2}$. In this case, the measurement amounts to guessing uniformly at random, which then succeeds with probability at least $1/2$. We will show that this success probability also applies when the inner loop is run. Note that the value for t_k is the same as $\frac{\pi}{4\alpha} - \frac{1}{2}$ rounded to the nearest integer, and the $-\frac{1}{2}$ part comes because we start with $\theta = \alpha$ not $\theta = 0$.

13.3 The Analysis

Let θ be the angle between $\boldsymbol{m}$ and the current state $\boldsymbol{a}$ before any iteration of the inner loop. Suppose $\alpha \leq \theta < \frac{\pi}{2}$; note that initially $\theta = \alpha$. Because $\boldsymbol{m}$ is our x-axis, the Grover reflection puts $\boldsymbol{a}'$ at angle $-\theta$, which is stepping back to spring forward. Because its distance from $\boldsymbol{j}$ is now $\alpha + \theta$, the reflection about $\boldsymbol{j}$ doubles that and adds it to $-\theta$. The new angle is hence

$$\theta' = -\theta + 2\alpha + 2\theta = \theta + 2\alpha,$$

which means that the two reflections have effected a positive rotation by 2α. The final angle θ will hence lie within $\pm\alpha$ of $\frac{\pi}{2}$. Because $\alpha \leq \frac{\pi}{4}$ for any iteration to happen at all, the success probability $\sin^2(\theta)$ is at least $1/2$.

Finally, the inequality $\sin(x) \leq x$ means $\alpha \geq \sqrt{\frac{k}{N}}$, so $t_k \approx \frac{\pi}{4\alpha} < \frac{1}{\alpha} \leq \sqrt{\frac{N}{k}}$. It follows that even when $k = 1$, we have proved:

THEOREM 13.1 Given a function $f\colon \{0,1\}^n \to \{0,1\}$ from a feasible family, and given $k = |S|$ where $S = \{x \mid f(x) = 1\}$, Grover's algorithm finds a member of S in an expected number $O(T)$ of iterations, where $T = 2^{(n-\log_2 k)/2}$, and in overall time $Tn^{O(1)}$. □

13.4 The General Case, with k Unknown

Now we consider what happens when k is not known in advance. If we operate as though $k = 1$, then we might overshoot because the rotation amount 2α depends on the actual value of k. We might be unlucky enough to land all the way on the other side of the circle at an angle near π or even back where we started, whereupon measuring would give success probability near zero.

Suppose instead that we try to be cautious and measure after every t-th iteration. Because measurement collapses the system, we would have to restart. The expected time then becomes the sum of t from 1 to t_k, which is of order t_k^2, and would exactly cancel the quadratic savings granted by the procedure when t_k is known. One idea would be to try to maintain multiple computations so there is always a spare copy each time we measure one. However, this could require preparing a huge number t_k of ancilla qubits, and insofar as maintaining them might involve keeping up a $t_k \times t_k$ grid, could likewise cancel the time savings.

The simplest of several known solutions is to choose the stopping time t for the iterations at random. The key is that the success probability p of measuring at the time $\boldsymbol{a}$ has angle θ is given by $\sin^2(\theta)$. This is like a sine wave except steeper and with period π not 2π staying in non-negative values. Nevertheless, if we throw a dart along the horizontal axis of a graph of $\sin^2(\theta)$ to choose θ at random, then there is exactly a 50-50 chance of it being more than 45 degrees away from the x-axis, which still gives $\sin^2(\theta) \geq 1/2$. We need only beware of choosing our range of dart values too narrowly when α is large, but we can guard against too-large α by making the classical guess-at-random step come first. The revised algorithm is:

1. Classically guess $x \in \{0,1\}^n$ uniformly at random. If success, i.e., $x \in S$, **stop**.
2. Initialize the vector $\boldsymbol{a}$ to be the start vector $\boldsymbol{j}$.
3. Randomly select a number t such that $1 \leq t \leq \sqrt{N}/2$.
4. Repeat t times:
   4.1 Apply $\mathbf{Ref}_m$ to $\boldsymbol{a}$ via the Grover oracle $\mathbf{U}_S$, obtaining the vector $\boldsymbol{a}'$.
   4.2 Apply $\mathbf{Ref}_j$ to $\boldsymbol{a}'$, obtaining the new value of $\boldsymbol{a}$.
5. Measure the final state $\boldsymbol{a}$, giving a string $x \in \{0,1\}^n$.
6. If $x \in S$ stop—we have found a solution. Otherwise repeat the entire process.

The analysis is straightforward. The first step succeeds with probability at least $\frac{1}{4}$ unless $\alpha < \frac{\pi}{6}$ because the sine of 30 degrees is $\frac{1}{2}$. Otherwise, one need only show that regardless of α such that $1/\sqrt{N} < \alpha < \frac{\pi}{6}$, a random rotation by $t \cdot 2\alpha$ has at least a $\frac{1}{2} - \epsilon$ chance of falling within $\frac{\pi}{4}$ of the y-axis, giving overall success probability at least $\frac{1}{2}(\frac{1}{2} - \epsilon)$, where ϵ is tiny in absolute terms. (The exercises ask how to arrange rigorously to get the probability over 24% for all n and possible values of k.)

13.5 Grover Approximate Counting

We can blend Grover's search with Shor's algorithm to estimate the number k of solutions. This is equivalent to estimating the angle of rotation 2α. For intuition, let us first suppose $2\alpha = 2\pi/r$ for some integer r. Then the function $f(t) = \sin^2(2t\alpha)$ is periodic with period r. Hence, we can apply Shor's algorithm to find the period r, which in this case would tell us k exactly, and the application adds only $n^{O(1)} = (\log N)^{O(1)}$ time. That $f(t)$ is not injective is OK because it is at worst 4-to-1, and the case where θ does not divide the circle evenly will still leave us able to approximate r and then estimate k. The issue is that if we followed chapter 11 literally, it would require first computing the functionally superposed state of $f(t)$ over all t, but at first it seems hard how to get $f(t)$ without sampling from repeated measurements.

The answer is that the quantum state of Grover's algorithm after t iterations, superposed over all integers t up to at most $\sqrt{N}$, already contains enough of the right kind of information to make Shor's algorithm work. The QFT will amplify the results for those t that are close to the optimal iteration number

$\tilde{t} \approx \sqrt{N/k}$. That is, if we measure the qubits that started off holding values of t, most of the amplitude will have been attracted to those t that are close to $\tilde{t}$. It will help if we can avoid "overshooting" by preventing $t > 2\tilde{t}$, which could cancel good results. We will structure the algorithm to provide a good chance of success before this could happen.

This conveys the essential idea, and at this point, it is fine to skip ahead to the next chapter on quantum random walks. The rest of this section is to make good on our intent to provide full details. It also exemplifies a *phase-estimation* task that is substantial but easier than the one needed for theorem 15.2 in chapter 15, where we do skip the proof details.

Our goal is to estimate k to within a factor of $(1+\epsilon)$, and we will succeed on pain of multiplying the expected time by $1/\epsilon$. Setting ϵ to be any inverse polynomial in n multiplies the time by only a polynomial in $\log N$, for overall time still $\tilde{O}(\sqrt{N})$. By the same measure, we can afford restarting Grover's algorithm $\log_2 \sqrt{N}$ times, each time guessing for the true $\tilde{t}$ to be double what we tried before. Note that when $t \sim 2^{\ell} \ll \tilde{t}$, this means we are guessing a k that is much higher than the true value, and then with high probability the Shor-based counting routine will say "zero"—whereupon we increment ℓ and try again.

For our estimates, we will need finer trigonometric analysis than we have used before. Because this is an advanced section, we refer proofs of the following two inequalities to general sources.

LEMMA 13.2 For any $M \geq 1$, $\delta > 0$, and angles θ, α, β with $0 \leq M\theta \leq \frac{\pi}{2}$ and $|\alpha - \beta| \leq \delta$:

$$\sin(M\theta) \geq M\sin(\theta)\cos(M\theta). \tag{13.1}$$

$$|\sin^2(\alpha) - \sin^2(\beta)| \leq 2\delta|\sin(\alpha)\cos(\alpha)| + \delta^2. \tag{13.2}$$

The distinctive feature of this algorithm is a superposition over t, using $\{0,1\}^{\ell}$ to code $[0, 2^{\ell}-1]$. We will begin with a $1/\sqrt{2^{\ell}}$-weighted superposition of states of the form

$$\boldsymbol{e}_t \otimes \boldsymbol{j}_n,$$

where on each one we intend t steps of Grover iteration starting with the $\boldsymbol{j}_n$ part. The issue—explored in problem 13.9 below—is that carrying along a dependent value in the first ℓ quantum coordinates upsets the geometry of the rotations. This applies not for the reflection about $\boldsymbol{m}$, which becomes just a sign flip, but for the one about $\boldsymbol{j_n}$ itself. That is, the two-dimensional space has to choose a fixed "*j*-vector," so it does projections based on $\boldsymbol{j}' = \boldsymbol{e}_{0^\ell} \otimes \boldsymbol{j_n}$. To use $\boldsymbol{j}'$, we must at each step *undo* the transformation used to create the kind of initial superposition over t, reflect about $\boldsymbol{j}'$, and then *redo* the transformation to re-create the functional superposition needed for the Grover oracle call to work on each superposed track where the iteration is still active. Because the QFT gives the same result as the Hadamard transform on the all-zero basis state, we can use the QFT as a partial transform on the first ℓ qubits to create the superposition. The iteration operator for $M = 2^\ell$ is

$$\boldsymbol{Q} = \boldsymbol{F}_M \boldsymbol{Ref}_{j'} \boldsymbol{F}_M^{-1} \boldsymbol{Ref}_{\boldsymbol{m}}.$$

Happily, the extra applications of $\boldsymbol{F}_M$ affect only the polylog-in-N factors in the cost. Now we need to iterate $\boldsymbol{Q}$ a different number of times for each t. This gives rise to the operation

$$\boldsymbol{Stagger}^{(2^\ell)}(\boldsymbol{Q})(\boldsymbol{e}_t \otimes \boldsymbol{j_n}) = \boldsymbol{Q}^t(\boldsymbol{e}_{0^\ell} \otimes \boldsymbol{j_n}).$$

This looks hard to perform, but the control tricks in chapter 6 show how. $\boldsymbol{Stagger}^{(M)}$ can be coded by treating the first $\log_2 M$ quantum coordinates as a counter initialized to one of the superposed t, which gets decremented with each iteration. An iteration uses controlled gates to perform $\boldsymbol{Q}$ conditioned on the counter not being zero. Everything works in superposition without making more than M Grover oracle calls overall.

The last detail is how to choose M. As with Grover's original algorithm in the case where k is known, we want to avoid doing too many iterations, but here the motive is different. Because we are superposing over all $t < M$, overshooting is not the main issue. We will show in the proof that once M is above a certain threshold, the value of M does not matter much to either the quality or likelihood of the estimate obtained. Remarkably, with probability over $\frac{8}{\pi^2} \geq 0.81$, the measurement will yield one of the two integers that flank the optimal fractional value. Instead, the motive is just to minimize the number of queries and the running time, keeping $M = O(\frac{1}{\epsilon})$. Because our k is unknown—indeed, k is exactly what we are trying to estimate—we do not know this threshold in advance, but we can "probe" for it by restarting with different values for ℓ until we are close enough that the returned estimate for k is nonzero. Then the final value of ℓ gives enough guidance on how far to jump M ahead for the final run.

13.5.1 The Algorithm

As in section 13.4, we can preface the algorithm with classical random sampling to catch the case of $\frac{k}{N}$ greater than $\frac{1}{4}$ or some smaller constant. Assuming the samples are all "misses," we proceed to the quantum part.

1. Choose ϵ based on n, N, and initialize $\ell = 1$, $M = 2^\ell$.
2. Apply $\boldsymbol{F}_M$ to the first part of the start vector $\boldsymbol{a}_0 = \boldsymbol{e}_{0^\ell} \otimes \boldsymbol{j}_n$ to get $\boldsymbol{a}$.
3. While ($M < \sqrt{N}$) do:
   3.1 Compute $\boldsymbol{a}' = \textbf{Stagger}^{(M)}(\boldsymbol{Q})\boldsymbol{a}$;
   3.2 Apply $\boldsymbol{F}_M^{-1}$ once more to the first ℓ qubits to make $\boldsymbol{a}''$;
   3.3 Measure $\boldsymbol{a}''$, reading the result v on the first ℓ qubits as a number in $0 \dots M-1$;
   3.4 If $v > 0$, then break; otherwise do $\ell = \ell + 1$, $M = 2^\ell$, $\boldsymbol{a} = \boldsymbol{a}_0$, and begin the next while-loop iteration.
4. Using the last value ℓ in the loop, set M to be the smallest power of 2 above $\frac{20\pi^2}{\epsilon} 2^\ell$.
5. Form $\boldsymbol{a}_0$ and apply $\boldsymbol{F}_M$ as a partial transform on the first $\ell' = \log_2 M$ qubits to get $\boldsymbol{a}$.
6. Repeat the while loop once through and measure to get the final value v.
7. Round $N \sin^2(\pi \frac{v}{M})$ to an integer—not caring up or down if the fractional part is near 0.5—and output it as the estimate k' for k.

13.5.2 The Analysis

First note that if $k = 0$, then $\boldsymbol{m}$ and $\boldsymbol{j}$ coincide, so that the sign-flip action commutes with $\boldsymbol{F}_M$. This causes everything to cancel, leaving $\boldsymbol{a}'' = \boldsymbol{a}_0$, whose measurement in the first ℓ qubits always gives $v = 0$. The only pain is that this takes the maximum while-loop time to find out, and in particular it makes the full budget of about $2\sqrt{N}$ Grover oracle queries. For $k > 0$, the time is tighter:

THEOREM 13.3 If $k > 0$, then with probability at least 2/3, the algorithm outputs k' such that $|k' - k| \leq \epsilon k$, while using $O(T)$ evaluations and $\tilde{O}(T)$ time overall, where

$$T = \frac{1}{\epsilon}\sqrt{\frac{N}{k}}.$$

Proof. With reference to section 13.3, let α be the angle in radians between $\boldsymbol{j}$ and $\boldsymbol{m}$, and set $m = \lfloor \log_2(\frac{1}{5\alpha}) \rfloor$. We first claim that with probability at least $\cos^2(2/5)$, the while-loop gives 0 for all $\ell \leq m$. We will then show that on the last stage with $\ell = m+1$, we get a good nonzero value with probability at least $8/\pi^2$. The conclusion will follow because $\cos^2(0.4) = 0.848\ldots$ and $8/\pi^2 = 0.81\ldots$, with product $> 2/3$. We will not need to consider the eventuality that the while-loop bound $\sqrt{N}$ is exceeded. Let

$$s_M(b) = \frac{1}{\sqrt{M}} \sum_{y=0}^{M-1} e^{2\pi iby} \boldsymbol{e}_y.$$

Then $\boldsymbol{F}_M \boldsymbol{e}_x = s_M(\frac{x}{M})$. It follows that if b is an integer multiple x of $\frac{1}{M}$, then measuring $\boldsymbol{F}_M^{-1} s_M(b)$ recovers $\boldsymbol{e}_x$ with certainty, even for $x = 0$. Moreover, for any b and x, the chance of obtaining $\boldsymbol{e}_x$ by measuring $\boldsymbol{F}_M^{-1} s_M(b)$ is

$$\begin{aligned}
p_x &= |\langle \boldsymbol{e}_x, \boldsymbol{F}_M^{-1} s_M(b) \rangle|^2 \\
&= |\langle \boldsymbol{F}_M^* \boldsymbol{e}_x, s_M(b) \rangle|^2 \\
&= |\langle s_M^*(\frac{x}{M}), s_M(b) \rangle|^2 \\
&= \left| \frac{1}{M} \left(\sum_{y=0}^{M-1} e^{-2\pi i \frac{x}{M} y} \boldsymbol{e}_y \right) \left(\sum_{y=0}^{M-1} e^{2\pi iby} \boldsymbol{e}_y \right) \right|^2 \\
&= \frac{1}{M^2} \left| \sum_{y=0}^{M-1} e^{2\pi idy} \right|^2,
\end{aligned}$$

writing $d = |\frac{x}{M} - b|$. When $\boldsymbol{Stagger}^{(M)}(Q)$ is applied to $(\boldsymbol{F}_M \boldsymbol{e}_{0^\ell}) \otimes \boldsymbol{j}_n$, we have $x = 0$ and $b = \frac{\alpha}{\pi}$, so $d = \frac{\alpha}{\pi}$. Thus, by the derivation of (11.2) in section 11.4, we finally obtain

$$p_0 = \frac{1}{M^2} \cdot \frac{\sin^2(M\pi d)}{\sin^2(\pi d)},$$

which is the chance of the measurement giving $\boldsymbol{e}_0$, that is, 0. Accordingly, the probability of getting the first m trials zero is

$$p_0 = \prod_{\ell=1}^{m} \frac{\sin^2(2^\ell \alpha)}{2^{2\ell} \sin^2(\alpha)}.$$

Because $2^\ell\alpha \le 2^m\alpha \le \frac{1}{5} < \frac{\pi}{2}$ satisfies the hypothesis of inequality (13.1) in lemma 13.2, we can apply $\sin(M\alpha) \ge M\sin(\alpha)\cos(M\alpha)$ to yield

$$p_0 \ge \prod_{\ell=1}^{m} \cos^2(2^\ell\alpha) = \frac{1}{2^{2m}} \cdot \frac{\sin^2(2^{m+1}\alpha)}{\sin^2(2\alpha)}.$$

Applying the inequality again with 2^m in place of M and 2α in place of α makes

$$p_0 \ge \frac{2^{2m}\sin^2(2\alpha)\cos^2(2^{m+1}\alpha)}{2^{2m}\sin^2(2\alpha)} = \cos^2(2^{m+1}\alpha) = \cos^2(\frac{2}{5}),$$

by the choice of m. This completes the first goal.

For the second goal, given that the first goal has succeeded, we note that because the loop has $\ell = m+1$ on the first nonzero value, the inequality $2^{m+1} > \frac{1}{5\alpha}$ gives

$$M \ge \frac{20\pi^2}{\epsilon}2^{m+1} \ge \frac{4\pi^2}{\epsilon\alpha}.$$

Using that $\alpha \le \frac{\pi}{2}\sin(\alpha) = \frac{\pi}{2}\sqrt{k/N}$, this further yields

$$M \ge \frac{8\pi}{\epsilon\sin(\alpha)} = \frac{8\pi\sqrt{N/k}}{\epsilon}.$$

Let $\tilde{v}$ be such that $\frac{k}{N} = \sin^2(\pi\frac{\tilde{v}}{M})$. For a result v that we get, we are interested in $|\sin^2(\pi\frac{v}{M}) - \sin^2(\pi\frac{\tilde{v}}{M})|$. When v is one of the two flanking integers of $\tilde{v}$, the difference in the angles will be at most $\delta = \frac{\pi}{M}$. By (13.2) in lemma 13.2, we will have:

$$\begin{aligned}|\sin^2(\pi\frac{v}{M}) - \sin^2(\pi\frac{\tilde{v}}{M})| &\le 2\delta|\sin(\pi\frac{\tilde{v}}{M})\cos(\pi\frac{\tilde{v}}{M})| + \delta^2\\ &= 2\delta\sqrt{\sin^2(\pi\frac{\tilde{v}}{M})(1-\sin^2(\pi\frac{\tilde{v}}{M}))} + \delta^2\\ &= 2\delta\frac{\sqrt{k(N-k)}}{N} + \delta^2.\end{aligned}$$

Substituting for δ gives:

$$\begin{aligned}2\pi\frac{\sqrt{k(N-k)}}{NM} + \frac{\pi^2}{M^2} &\le 2\pi\frac{\sqrt{k(N-k)}}{N\cdot 8\pi\sqrt{N/k}/\epsilon} + \frac{\pi^2}{64\pi^2N/k\epsilon^2}\\ &= \frac{\epsilon}{4N}\sqrt{k^2\left(\frac{N-k}{N}\right)} + \frac{k\epsilon^2}{64N}\\ &\le \frac{\epsilon}{4N}k + \frac{\epsilon^2}{64N}k.\end{aligned}$$

Hence, the additive error in the estimate $N\sin^2(\pi\frac{v}{M})$ before rounding is at most $k\left(\frac{\epsilon}{4}+\frac{\epsilon^2}{64}\right)$. This is small enough that the k' obtained after rounding gives

$$|k-k'|\leq\epsilon k,$$

which is what we need to prove. Hence, we have done everything except lower-bound the probability of getting the measured v to be one of the two flanking integers of $\tilde{v}$.

To do this, we return to the geometry of what is happening in our product Hilbert space of the first ℓ' qubits and the space spanned by the original Grover hit and miss vectors $\boldsymbol{h}$ and $\boldsymbol{m}$ on the final run-through from step 5. The vector $\boldsymbol{a}_0$ in this representation is

$$\boldsymbol{a}_0=\frac{-i}{\sqrt{2}}\boldsymbol{e}_0\otimes(e^{i\alpha}\boldsymbol{h}-e^{-i\alpha}\boldsymbol{m}).$$

After the application of $\boldsymbol{F}_M$ as a partial transform on the first part, and ignoring the global phase factor that came from the rotated basis, we have

$$\boldsymbol{a}=\frac{1}{\sqrt{2M}}\sum_{y=0}^{M-1}\boldsymbol{e}_y\otimes(e^{i\alpha}\boldsymbol{h}-e^{-i\alpha}\boldsymbol{m}).$$

After the differential numbers of Grover rotations given by $\boldsymbol{Stagger}^{(M)}(\boldsymbol{Q})$, we have:

$$\begin{aligned}\boldsymbol{a}'&=\frac{1}{2M}\sum_{y=0}^{M-1}\boldsymbol{e}_y\otimes(e^{i(\alpha+2y\alpha)}\boldsymbol{h}-e^{-i(\alpha+2y\alpha)}\boldsymbol{m})\\&=\frac{e^{i\alpha}}{\sqrt{2M}}\sum_{y=0}^{M-1}e^{i2y\alpha}(\boldsymbol{e}_y\otimes\boldsymbol{h})-\frac{e^{-i\alpha}}{\sqrt{2M}}\sum_{y=0}^{M-1}e^{-i2y\alpha}(\boldsymbol{e}_y\otimes\boldsymbol{m})\\&=\frac{e^{i\alpha}}{\sqrt{2}}\boldsymbol{s}_M(\frac{\alpha}{\pi})\otimes\boldsymbol{h}-\frac{e^{-i\alpha}}{\sqrt{2}}\boldsymbol{s}_M(\frac{\pi-\alpha}{\pi})\otimes\boldsymbol{m}.\end{aligned}$$

Hence, what we measure in the end after applying the final $\boldsymbol{F}_M^{-1}$ on the first space is an evenly weighted mix of measuring either $\boldsymbol{F}_M^{-1}\boldsymbol{s}_M(\frac{\alpha}{\pi})$ or $\boldsymbol{F}_M^{-1}\boldsymbol{s}_M(\frac{\pi-\alpha}{\pi})$. Now the amplitude the latter contributes to $\boldsymbol{e}_y$ equals what the former contributes to $\boldsymbol{e}_{M-y}$, and because

$$\sin^2(\pi\frac{M-y}{M})=\sin^2(\pi\frac{y}{M}),$$

the overall probability of obtaining y is the same as what we get from measuring $\boldsymbol{F}_M^{-1} s_M(\frac{\alpha}{\pi})$. If $\frac{\alpha}{\pi}$ were an integer multiple y of $1/M$, then by the Fourier analysis at the start of this proof, we would obtain y with certainty. As it is, we need only lower-bound the probability of getting y such that $|\frac{y}{M} - \frac{\alpha}{\pi}| \leq \frac{1}{M}$. Let $D = M\frac{\alpha}{\pi} - \lfloor M\frac{\alpha}{\pi} \rfloor$, so $0 \leq D < 1$, and $d = \frac{D}{M}$. Then we have that the probability of the measurement y producing either the integer above or the integer below the target $\tilde{v}$ is

$$\frac{\sin^2(M\pi d)}{M^2 \sin^2(\pi d)} + \frac{\sin^2(M\pi(\frac{1}{M} - d))}{M^2 \sin^2(\pi(\frac{1}{M} - d))}.$$

This attains its minimum for $d = \frac{1}{2M}$, uniquely when $M > 2$, whereupon it becomes

$$2\frac{\sin^2(\frac{\pi}{2})}{M^2 \sin^2(\frac{\pi}{2M})} \geq 2\frac{1}{M^2}(\frac{2M}{\pi})^2 = \frac{8}{\pi^2}.$$

Thus, with at least this probability, the value $y = v$ returned by the measurement gives $|\pi\frac{v}{M} - \alpha| \leq \frac{\pi}{M}$, and it follows that the estimate $k' = N\sin^2(\pi\frac{v}{M})$ is within ϵk of the true value k. □

The analysis still allows bad values with probability slightly less than 1/3. However, the displacement of success away from 1/2 implies that with repeated trials, there will be clustering around a unique correct value, and averaging the cluster (while discriminating away outliers) will produce an estimate with higher confidence.

13.6 Problems

13.1. Calculate the 2×2 matrix of the action of one iteration $\boldsymbol{Ref}_j \boldsymbol{Ref}_m$ in the $\boldsymbol{h}, \boldsymbol{m}$ basis. Here the first dimension holds the aggregate value of the k-many coordinates that are hits, and the other holds the value of the $N - k$ misses. Note that the hit vector $\boldsymbol{h}$ becomes $(1, 0)$, and $\boldsymbol{m}$ becomes $(0, 1)$.

13.2. With reference to problem 13.1, what does $\boldsymbol{j}$ become as a unit vector in the $\boldsymbol{h}, \boldsymbol{m}$ basis? What does the action look like when the 2×2 matrix is given with respect to the (nonorthogonal) basis formed by $\boldsymbol{h}$ and $\boldsymbol{j}$ instead?

13.3. For what initial value of θ does Grover's algorithm guarantee finding a solution (100% probability) with a measurement after one iteration? Find the corresponding value of k.

13.4. What happens if you do another Grover iteration when $\theta = \frac{\pi}{4}$ (i.e., 45°) rather than stop as the algorithm indicates?

13.5. Calculate the exact number of Grover iterations when $N = 2^8 = 256$ and $k = 4$. Is it the same as when $N = 64$ and $k = 1$? What is the success probability of one trial in each case?

13.6. Complete the analysis of the general case in section 13.4, choosing t and bounding α more carefully to make it work with $\epsilon = 0.01$, giving success probability at least 0.24 in each trial.

13.7. Show how to use Grover's algorithm to decide whether an n-vertex graph has a triangle in quantum time about $O(n^{3/2})$.

13.8. Consider the following alternative to the strategy of section 13.4 for the case where the number k of solutions is unknown. For some integer $c > 0$, first do $c+1$ trials for $t = 1$ to $c+1$ in which measurement is done after t iterations, restarting the whole process if a solution is not found. Then obtain the next t' by multiplying the old t by $(1 + \frac{1}{c})$ and rounding up so the sequence continues $c+3, c+5, \ldots, 2c+1, 2c+4, \ldots$, eventually growing exponentially. Give an estimate for the expected running time as a function of k and n. What value of c minimizes your estimate?

13.9. Suppose we want to execute Grover search with functionally superposed states. That is, for some (classically feasible) function g, whenever the basic Grover algorithm is in a state $\sum_x a_x \mathbf{e}_x$, our algorithm will need to be in the state

$$\sum_x a_x |x\rangle |g(x)\rangle = \sum_x a_x \mathbf{e}_x \otimes \mathbf{e}_{g(x)}.$$

Show that the algorithm works unchanged provided $g(x)$ is a constant function.

13.10. When g is not constant, does the idea in problem 13.9 work?

13.11. Suppose $S(n)$ is the time to compute

$$G(\mathbf{e}_{x0^m}) = \mathbf{e}_{xg(x)}$$

whether on a single basic input x or a superposed one. Show that we can modify the Grover iteration to work with the functional states, on pain of $2S(n)$ becoming an extra multiplicative factor on the time. (Replicating this "setup time" for the functional superposition was OK in section 13.5 but will not be in chapter 15.)

13.7 Summary and Notes

Grover's algorithm appeared in the 1996 STOC conference (Grover, 1996) and the next year in full journal form (Grover, 1997). Originally, it assumed that there is only one solution. Later it was realized that the idea works for any number of solutions. Unlike the case of Shor's algorithm, there has been relatively little practical argument against the assertion that the functionally superposed states $\frac{1}{\sqrt{N}} \sum_x |x\rangle|f(x)\rangle$ are feasible to prepare, so that the oracle operation $\boldsymbol{U}_f$ is feasible. It has been noted, however—see discussion in the texts by Hirvensalo (2010) and Rieffel and Polak (2011)—that for many particular functions f, similar speedup can be obtained by classical means. This leaves the question of building $\boldsymbol{U}_f$ for functions f for which sharp lower bounds on classical search might be provable.

Another question, especially when k is known, is why can't we jump ahead by computing the right number of iterations in one sweep? The answer is that computation apart from $\boldsymbol{U}_f$ can gain only no or little information about the angle of the hit vector, whereas $\boldsymbol{U}_f$ provides only the given small angle on each call. It has been proven rigorously in various ways—see Bennett et al. (1997) and Beals et al. (2001)—that $\Omega(\sqrt{\frac{N}{k}})$ calls to $\boldsymbol{U}_f$ are necessary, making this an asymptotically tight bound up to constant factors. Our long section on approximate counting follows the main article (Brassard et al., 2000), which gives full details over the earlier conference version (Brassard et al., 1998). The former article also gives more general applications, trade-offs between success probability and closeness of the estimate, and some results with nontrivial exact counting. Problem 13.8 follows lecture notes (https://cs.uwaterloo.ca/~watrous/LectureNotes.html) by John Watrous.

Grover's algorithm is greatly important for two reasons. First, it gives only a polynomial speedup: a search of cost T roughly becomes a search of cost $\sqrt{T}$. But it is completely general. This ability to speed up almost any kind of search has led to a large amount of research, even more than what we exemplify in chapters 14 and 15. Perhaps if quantum computers one day are real, Grover's algorithm may be used in many ways in practice.

Second, it showed that there were other types of algorithms. All the previous algorithms that we have discussed had just a few steps and/or turned on an immediate property of the Hadamard or Fourier transform. Grover's algorithm did the most to break this form and pointed the way to more intricate algorithms when combined with quantum walks, which we cover next.

14 Quantum Walks

This chapter gives a self-contained and elementary presentation of quantum walks, needing only previous coverage in this text of graph theory and the sum-over-paths behavior. In the next chapter, where we cover search algorithms using quantum walks, the material is necessarily more advanced, and we have chosen to emphasize the intuition at some expense of detail.

Both quantum and classical random walks can be visualized as walks on graphs. The graphs may be finite or infinite, directed or undirected. We will work toward a bird's-eye view of a *quantum walk* as a deterministic process (before any measurement) and will follow recent usage of excising the word "random" in the quantum case. First we consider the classical case.

14.1 Classical Random Walks

Classical random walks on graphs are a fundamental topic in computational theory. The idea of a walk is easy to picture. Suppose you are at a node $u \in V$, and suppose there are edges out of u to neighbors $v_1, \dots, v_d$. In the **standard random walk**, you pick a neighbor v_i at random, that is, with probability $1/d$. In a general random walk, there is a specified probability $p_{u \to v_i}$ for each neighbor v_i. Either way, if v_i is chosen, then you go there by setting $u = v_i$ and repeat this process. There are three main questions about a classical random walk:

1. Given a node w different from u, what is the expectation for the number of steps to reach w starting from u?
2. How many steps are expected for the walk to visit all nodes w in the graph, in case $n = |V|$ is finite?
3. If you stop the walk after a given number t of steps, what is the probability $p_t(w)$ of ending at node w? How does it behave as t gets large?

The questions can have dramatically different answers depending on whether G is directed or undirected. To see this, first consider the undirected graph in which the vertices stand for integers i, and each i is connected to $i-1$ and $i+1$. If we start at 0, then what is the expected number of steps to reach node n? Each step is a coin flip—heads you move right, tails you move left. Hence, reaching cell n means sometimes having an excess of n more heads than tails. Now the standard deviation of N-many coin flips is proportional to $\sqrt{N}$, and it follows that the expected time to have a positive deviation of n is $O(n^2)$.

This result carries over to any undirected n-vertex graph. If node y is reachable at all from node x, then there is a path from x to y of length at most $n-1$.

It is possible that some node u along this path may have degree $d \geq 3$ with $d-1$ of the neighbors farther away from y, so that the chance of immediate progress is only $1/d$. However, this entails that the original distance from x to y was at most $n-d+1$. Thus, any graph structure richer than the simple path trades against the length, and it can be shown that the $O(n^2)$ step expectation of the simple path remains the worst case to reach any given node in the same connected component as x. In particular, this yields an economical randomized algorithm to tell whether an undirected graph is connected by taking a walk of $O(n^2)$ steps and tracking the different nodes encountered.

For directed graphs, however, the time can be exponential. Consider directed graphs G_n with $V = \{0,\ldots,n-1\}$ and edges $(i,i+1)$ and $(i,0)$ for each i. The walk starts at $u = 0$ and has goal node $y = n-1$, which we may suppose has both out-edges going to 0. Now a "tail" sends the walk all the way back to 0, so the event of reaching y is the same as getting $n-1$ consecutive heads. The expected time for this is proportional to 2^n. Thus, mazes with one-way corridors are harder to traverse than the familiar kind with undirected corridors.

There are two main further insights on the road to quantum walks. The first is that quite apart from how directedness can make locations difficult to reach with high probability, it is possible to cancel the probability of being in certain locations at certain times altogether. The second is like the difference between AC and DC electricity. Instead of seeing a walk as "going somewhere" like a direct current, it is better to view it as a dance back and forth on the vertices according to some eventually realized distribution. Both insights require representing walks in terms of actions by matrices, and again we can get much initial mileage from the classical case.

14.2 Random Walks and Matrices

Classical random walks on graphs $G = (V,E)$ can be specified by matrices $\boldsymbol{A}$ whose rows and columns correspond to nodes u,v. Here $\boldsymbol{A}$ is like the adjacency matrix of G, in that $\boldsymbol{A}[u,v] \neq 0$ only if there is an edge from u to v in G, but the entries on edges are probabilities. Namely, $\boldsymbol{A}[u,v] = p_{u \to v}$, which denotes the probability of going next to v if the "walker" is at u. The matrix $\boldsymbol{A}$ is row-stochastic as defined in section 3.5; that is, the values in each row are nonnegative and sum to 1.

It follows that $\boldsymbol{A}^2$ is also a row-stochastic matrix and gives the probabilities of pairs of steps at a time. That is, for any nodes u and w,

$$\boldsymbol{A}^2[u,w] = \sum_v \boldsymbol{A}[u,v]\boldsymbol{A}[v,w] = \sum_v p_{u\to v}p_{v\to w}.$$

Because the events of the walk going from u to different nodes v are mutually exclusive and collectively exhaustive, this sum indeed gives the probability of going from u to w in two steps. The same goes for $\boldsymbol{A}^3$ and paths of three steps, and so on for $\boldsymbol{A}^k$, all $k \geq 0$.

A probability distribution D on the nodes of G is **stable** under $\boldsymbol{A}$ if for all nodes v,

$$D(v) = \sum_u D(u)\boldsymbol{A}(u,v).$$

Intuitively, this says that if $D(v)$ is the probability of finding a missing parachute jumper at any location v, then the probability is the same even if the jumper has had time to do a random step according to $\boldsymbol{A}$ after landing. Mathematically, this says that D is an *eigenvector* of $\boldsymbol{A}$, with eigenvalue 1; the eigenvector is on the left, giving $D\boldsymbol{A} = D$.

If G is a connected, finite, undirected graph that is not bipartite, there is an integer k such that for all $\ell \geq k$ and all $x, y \in V$, there is a path of exactly ℓ steps from x to y. It follows that for for any matrix $\boldsymbol{A}$ defining a random walk on G, all entries of $\boldsymbol{A}^k$ and all higher powers are nonzero. It then further follows—this is a hard theorem—that the powers of $\boldsymbol{A}$ converge pointwise to a matrix $\boldsymbol{A}^*$ that projects onto some stationary distribution. That is, for any initial distribution C, $C\boldsymbol{A}^* = D$, and moreover the sequence $C_k = C\boldsymbol{A}^k$ converges pointwise to D. This goes even for the distribution $C(u) = 1$, $C(v) = 0$ for all $v \neq u$, which represents our random-traveler initially on node u.

When $\boldsymbol{A}$ is the standard random walk, the limiting probability is

$$D(u) = \frac{deg(u)}{2|E|}.$$

Nonuniform walks $\boldsymbol{A}$ may have other limiting probabilities, but they still have the remarkable property that any initial distribution is converged pointwise to D. The relation between $\epsilon > 0$ and the power k needed to ensure $||C\boldsymbol{A}^\ell - D|| \leq \epsilon$ for all $\ell \geq k$, where $||\cdot||$ is the max-norm, is called the **mixing time** of $\boldsymbol{A}$, while the k giving $\max_{u,v} |D(v) - \boldsymbol{A}^\ell[u,v]| \leq \epsilon$ for all $\ell \geq k$ is called the **hitting time**.

If G is bipartite, there is still a stationary D, but not all C will be carried toward it—any distribution with support confined to one of the two partitions will alternate between the partitions. When G is directed, similar behavior occurs with period 3 in a directed triangle, and so on. However, provided that for every u, v and prime p there is a path from u to v whose number of steps is not a multiple of p, the above limiting properties *do* hold for every random walk on G, and the notions of mixing and hitting times are well defined. In an undirected graph, for constant ϵ, the hitting time of any walk is polynomial, but in a directed graph, even the standard walk may need exponential time, as the directed graphs in the last section show.

The analogy here is that the stationary distribution D is like "AC current" in that you picture a one-step dance back and forth, but the overall state remains the same. This differs from the "DC" view of a traveler going on a random walk. What distinguishes the quantum case is that via the magic of quantum cancellation, we can often arrange for $D(v)$ to be *zero* for many undesired locations v and, hence, pump up the probability of the "traveler" being *measured* as being at a desired location u.

14.3 An Encoding Nicety

To prepare for the notion of quantum walks, we consider the probabilities p as being derived from a set C of random outcomes. In the background is a function $h(u, c) = v$ that specifies the destination node for each outcome c when the traveler is at node u.

To encode the standard random walk in which the next node is chosen with equal probability among all out-neighbors v of u, we simply take $|C|$ to be the least common multiple of the out-degrees of all the vertices in the graph. Then for each vertex, we assign outcomes in C to choices of neighbor evenly. This is well defined also for classes of infinite graphs of bounded degree. Indeed, the infinite path graph remains a featured example, taking $C = \{0, 1\}$ and thinking of c as a "coin flip." We could extend this formalism to allow arbitrary distributions on C, but uniform suffices for the main facts and applications.

Now we make a matrix $\boldsymbol{A}'$ whose rows index pairs u, c of nodes and random outcomes. We can write this pair without the comma. So we define

$$\boldsymbol{A}'[uc, v] = \begin{cases} 1 & \text{if } h(u, c) = v \\ 0 & \text{otherwise.} \end{cases}$$

Now each row has one 1 and $n-1$ 0s. We can, however, obtain the stochastic matrix $\boldsymbol{A}$ above via

$$\boldsymbol{A}[u,v] = \frac{1}{|C|}\sum_c \boldsymbol{A}'[uc,v].$$

If we had a nonuniform distribution on C, then we could use a weighted sum accordingly. Note also that our functional view of matrices makes this undisturbed by the possibility that V, and hence $\boldsymbol{A}$ and $\boldsymbol{A}'$, could be infinite.

We do one more notational change that already helps with the classical case by making the matrix square again. We make the same random outcome part of the column value as well by defining:

$$\boldsymbol{B}[uc,vc'] = \begin{cases} 1 & \text{if } h(u,c) = v \text{ and } c' = c, \\ 0 & \text{otherwise.} \end{cases}$$

Then $\boldsymbol{B}$ acts like the identity on the C-coordinates and like $\boldsymbol{A}'$ on the V-coordinates, that is, on the nodes. Now the stochastic matrix $\boldsymbol{A}$ is given by

$$\boldsymbol{A}[u,v] = \frac{1}{|C|}\sum_c \boldsymbol{B}[uc,vc].$$

The sum on the right-hand side goes down the diagonal of the C-part, much like the trace operation does on an entire matrix. It is called a *partial trace* operation, and is related to the ideas in section 6.7. In the classical case, all this trick does is get us back to our original idea of entries $\boldsymbol{A}[u,v]$ being probabilities, but it will help in the quantum case where they are *amplitudes*. Having digested this, we can progress to define quantum walks with a minimum of fuss.

14.4 Defining Quantum Walks

The reason we need the added notation of C in the quantum case is that on the whole space $V(G) \otimes C$, quantum evolution is an entirely deterministic process. What gives the effect of a randomized walk is that the action on C is unknown and unseen before being implicitly "traced out."

DEFINITION 14.1 A **quantum walk** on the graph G is defined by a matrix $\boldsymbol{U}$ with analogous notation to B above, but where $\boldsymbol{U}$ is unitary, and allowing the action $\boldsymbol{U}_C$ of $\boldsymbol{U}$ on the C coordinates to be different from the identity.

Indeed, in the case $|C| = 2$, by making $\boldsymbol{U}_C$ have the action of a 2×2 Hadamard matrix $\boldsymbol{H}$, we can also simulate the action of flipping the coin at each step. Again the action of $\boldsymbol{H}$ is deterministic, but because measurement involves making a choice over the entries of $\boldsymbol{H}$, the end effect is nondeterministic. Here is an example that packs a surprise.

Let G be the path graph with seven nodes, labeled $u = -3, -2, -1, 0, 1, 2, 3$. Our state space is $V(G) \otimes \{0, 1\}$. To flip a coin, we apply the unitary matrix $\boldsymbol{C} = \boldsymbol{I} \otimes \boldsymbol{H}$, where I is the seven-dimensional identity matrix. To effect the outcome b, we apply the 14×14 permutation matrix $\boldsymbol{P}$ that maps $(u, 0)$ to $(u-1, 0)$ and $(u, 1)$ to $(u+1, 1)$. Because we will apply this only three times to a traveler beginning at 0, it doesn't matter in figure 14.1 where $(-3, 0)$ and $(3, 1)$ are mapped—to preserve the permutation property they can go to $(3, 0)$ and $(-3, 1)$, respectively, thus making the action on $V(G)$ circular. Our walk matrix is thus $\boldsymbol{A} = \boldsymbol{PC}$. We apply $\boldsymbol{A}^3$ to the quantum basis state $\boldsymbol{a}_0$ that has a 1 in the coordinate for $(0, 0)$ and a 0 everywhere else.

Figure 14.1
Expanded graph G' of quantum walk on path graph G.

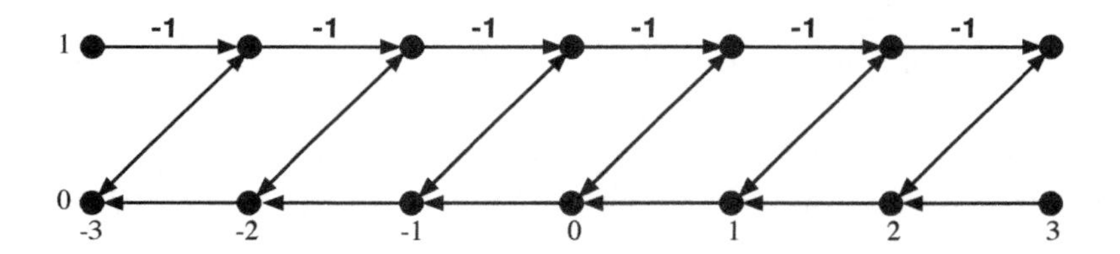

In three steps of a classical random walk on G starting at the origin, the probabilities on the nodes $(-3, -1, 1, 3)$, respectively, are $(\frac{1}{8}, \frac{3}{8}, \frac{3}{8}, \frac{1}{8})$ according to the familiar binomial distribution. (Those on the even nodes are zero because G is bipartite.) This is arrived at by summing over paths, each path being a product of three entries of the walk matrix. Because each nonzero entry is $\frac{1}{2}$, the middle values come about because there are three different ways to go from 0 to +1 in three steps and likewise from 0 to −1.

14.5 Interference and Diffusion

Having created the graph for the quantum walk, we need to say who our walker is going to be. Of course, it is the Feynman mouse Phil we encountered in chapter 7. That is to say, our quantum walk involves a sum over paths, with each path being a product of three entries in the matrix $\boldsymbol{A}$. Much is the same as with the three-stage mazes in that chapter. There is Hadamard cheese: the numerators of the nonzero entries can be -1 and $+1$. The denominators have $\sqrt{2}$ rather than 2 as with classical probability, but they will be squared when going from amplitudes to probabilities at the end. Third, and the unseen part under the hood, the paths being summed by nature fork not only in the G part but also in the C part of the space. That is to say, each coin outcome, which is represented by a column of the ordinary 2×2 Hadamard matrix,

$$\boldsymbol{H} = \begin{bmatrix} 1 & 1 \\ 1 & -1 \end{bmatrix},$$

has two ways of reaching that outcome, via the first or second row. When the coin outcome is 0 for "tails," both entries contribute a numerator of $+1$, but when the outcome is 1, one path contributes a $+1$ and the other -1.

Hence, the walk is really taking place in a 14-node graph G' that includes the coin flips. This graph has *directed* edges from $(u,0)$ and $(u,1)$ to $(u-1,0)$ for the outcome "tails," say wrapping around to $(3,0)$ in the case $u=-3$. For "heads," it has edges from $(u,0)$ and $(u,1)$ to $(u+1,1)$, again wrapping around, with the crucial difference that the rightward edges from $(u,1)$ (representing a previous outcome of heads) have multiplier -1. The other edges have $+1$. Now the three-step paths from $(0,0)$ in G', and their multiplier values, are:

$$\begin{array}{lllllllllll}
(0,0) & \to & (1,1) & \to & (2,1) & \to & (3,1) & : & 1\cdot -1 \cdot -1 & = & 1 \\
(0,0) & \to & (1,1) & \to & (2,1) & \to & (1,0) & : & 1\cdot -1 \cdot 1 & = & -1 \\
(0,0) & \to & (1,1) & \to & (0,0) & \to & (1,1) & : & 1\cdot 1 \cdot 1 & = & 1 \\
(0,0) & \to & (1,1) & \to & (0,0) & \to & (-1,0) & : & 1 & & \\
(0,0) & \to & (-1,0) & \to & (0,1) & \to & (1,1) & : & 1\cdot 1 \cdot -1 & = & -1 \\
(0,0) & \to & (-1,0) & \to & (0,1) & \to & (-1,0) & : & 1 & & \\
(0,0) & \to & (-1,0) & \to & (-2,0) & \to & (-1,1) & : & 1 & & \\
(0,0) & \to & (-1,0) & \to & (-2,0) & \to & (-3,0) & : & 1\,. & &
\end{array}$$

Thus, there are six, not four, different destinations. The crux is that the two paths that reach destination $(1,1)$ have multipliers of 1 and -1 and hence *cancel*, while the two paths with destination $(-1,0)$ both have multipliers of 1 and hence *amplify*. The 14-dimensional vector representing the quantum state of the outcome of the walk, given the initial vector that had a 1 in place $(0,0)$ and nothing else, becomes this when arranged in a 2×7 grid:

$$\boldsymbol{b}_0 = \frac{1}{\sqrt{8}} \begin{bmatrix} 0 & 0 & 1 & 0 & 0 & 0 & 1 \\ 1 & 0 & 2 & 0 & -1 & 0 & 0 \end{bmatrix}.$$

Finally, to obtain the probabilities while "tracing out" the unseen coin, we sum the *squares* in each column. The final classical probability vector, giving the probability of finding the "traveler" at each of the original seven nodes of G after a measurement, is

$$\left[\frac{1}{8}, 0, \frac{5}{8}, 0, \frac{1}{8}, 0, \frac{1}{8}\right].$$

This outcome of **diffusion** stands in marked contrast to the classical distributional outcome. What happened, and why the loss of symmetry?

It is at least reassuring that the bipartiteness of G showed through, giving zero probability again on the even-numbered vertices. But the leftward bias flies in the face of fairness when flipping a coin. The Hadamard matrix is used all the time to introduce quantum nondeterminism, so why does it give off-center results? The reason is that the -1 entry biases against "heads," causing cancellations that do not happen for "tails."

These cancellations can be harnessed for a rightward bias by starting the traveler at $(0,1)$ rather than $(0,0)$. That is, unknown to the traveler or any parties observing just the original graph G, we are starting the coin in an initial state of "heads" rather than "tails." In the $G \otimes C$ space, this means the initial state $\boldsymbol{a}_1$ has a 1 in the coordinate for $(0,1)$ and a 0 everywhere else. From $(0,1)$, some relevant three-step paths are:

$$\begin{array}{ccccccccccc} (0,1) & \rightarrow & (1,1) & \rightarrow & (2,1) & \rightarrow & (1,0) & : & -1 \cdot -1 \cdot 1 & = & 1 \\ (0,1) & \rightarrow & (1,1) & \rightarrow & (0,0) & \rightarrow & (1,1) & : & -1 \cdot 1 \cdot 1 & = & -1 \\ (0,1) & \rightarrow & (-1,0) & \rightarrow & (0,1) & \rightarrow & (1,1) & : & 1 \cdot 1 \cdot -1 & = & -1 \\ (0,1) & \rightarrow & (-1,0) & \rightarrow & (0,1) & \rightarrow & (-1,0) & : & 1 \cdot 1 \cdot 1 & = & 1 \\ (0,1) & \rightarrow & (1,1) & \rightarrow & (0,0) & \rightarrow & (-1,0) & : & -1 \cdot 1 \cdot 1 & = & -1. \end{array}$$

These show a mirror-image amplification and cancellation and give the 14-vector

$$\boldsymbol{b}_1 = \frac{1}{\sqrt{8}} \begin{bmatrix} 0 & 0 & 1 & 0 & -2 & 0 & -1 \\ 1 & 0 & 0 & 0 & 1 & 0 & 0 \end{bmatrix}.$$

These amplitudes give the classical probabilities $[\frac{1}{8}, 0, \frac{1}{8}, 0, \frac{5}{8}, 0, \frac{1}{8}]$ on G, now biased to the right.

Hence, you might think that you can cancel the biases by starting the system up with the coin in the "half-tails, half-heads" state

$$\boldsymbol{a}_2 = \frac{1}{\sqrt{2}}(\boldsymbol{a}_0 + \boldsymbol{a}_1).$$

This is like saying Schrödinger's cat has half a tail—or rather the square root of half a tail. By the linearity of quantum mechanics—remember $\boldsymbol{A}^3$ is just a matrix—the final state you get now is

$$\begin{aligned} \boldsymbol{b}_2 &= \frac{1}{\sqrt{2}}(\boldsymbol{b}_0 + \boldsymbol{b}_1) \\ &= \frac{1}{4}\begin{bmatrix} 0 & 0 & 2 & 0 & -2 & 0 & 0 \\ 2 & 0 & 2 & 0 & 0 & 0 & 0 \end{bmatrix}. \end{aligned}$$

Note that we got some more cancellations—that is to say, the two walks *interfered* with each other—and those were both on the right-hand side, so we have bias to the left again with probabilities $[\frac{1}{4}, 0, \frac{1}{2}, 0, \frac{1}{4}, 0, 0]$. In particular, the rightmost node is now unreached.

We can finally fix the bias by making the second walk occur with a quarter-turn *phase* displacement. This means starting up in the state

$$\boldsymbol{a}_3 = \frac{1}{\sqrt{2}}(\boldsymbol{a}_0 + i\boldsymbol{a}_1).$$

This state is like Schrödinger's cat with half a tail and an imaginary head. Again by linearity, the final state is $\boldsymbol{b}_3 = (\boldsymbol{b}_0 + i\boldsymbol{b}_1)/\sqrt{2}$, so that

$$\boldsymbol{b}_3 = \frac{1}{4}\begin{bmatrix} 0 & 0 & 1+i & 0 & -2i & 0 & 1-i \\ 1+i & 0 & 2 & 0 & -1+i & 0 & 0 \end{bmatrix}.$$

Taking the squared norms and adding each column gives the probabilities

$$\frac{1}{16}[0+2, 0, 2+4, 0, 4+2, 0, 2+0] = \left[\frac{1}{8}, \frac{3}{8}, \frac{3}{8}, \frac{1}{8}\right]$$

again, at last modeling the classical random walk.

A more robust way to fix the bias is to use a suitably "balanced" matrix other than Hadamard for the quantum coin-flip action. A suitable unitary matrix is

$$\boldsymbol{J} = \frac{1}{\sqrt{2}}\begin{bmatrix} 1 & i \\ i & 1 \end{bmatrix}.$$

If we take higher powers $\boldsymbol{A}^k$, whether $\boldsymbol{A} = \boldsymbol{P}(\boldsymbol{I} \otimes \boldsymbol{H})$ or $\boldsymbol{A} = \boldsymbol{P}(\boldsymbol{I} \otimes \boldsymbol{J})$ or whatever, then the circular connectivity of G guarantees that the uniform classical distribution giving probability $\frac{1}{7}$ to each node of G is approached. It is stable *as* the induced classical distribution.

14.6 The Big Factor

Now if we extend the graph G beyond seven nodes, then we come immediately to the surprise of greatest import. Let G have $n = 9$ nodes, labeled −4 to 4, and define G' as before including wrapping at the endpoints. Extend $\boldsymbol{P}$ to be 18×18 accordingly, keep $\boldsymbol{H}$ as the action on the coin space, and consider walks of length 4. Labeling the 16 basic walks of length 4 by HHHH through TTTT gives us a shortcut to compute the destination (m,a) and multiplier $b \in \{1,-1\}$ for each walk w:

- m is the number of H in w minus the number of T.
- a is 0 if w ends in T, and 1 if w ends in H.
- b is −1 if HH occurs an odd number of times as a substring of w, and 1 if even.

For example, HHTT, HTHT, and THHT all come back to $(0,0)$, and their respective multipliers are −1, 1, and −1. Importantly, this implies there is a cancellation at the origin, leaving −1 there. Similar happens with HTTH, THTH, and TTHH ending at $(0,1)$, leaving 1, whereas HTTT, THTT, and TTHT all hit $(-2,0)$ with weight 1, reinforcing each other to leave 3 there. The full 2×9 vector for the quantum state after the walk is:

$$\boldsymbol{b}_0 = \frac{1}{4}\begin{bmatrix} 0 & 0 & 1 & 0 & 1 & 0 & -1 & 0 & -1 \\ 1 & 0 & 3 & 0 & -1 & 0 & 1 & 0 & 0 \end{bmatrix}.$$

For the initial state $(0,1)$, we have the same rules, except with Hw in place of w. This yields

$$\boldsymbol{b}_1 = \frac{1}{4}\begin{bmatrix} 0 & 0 & 1 & 0 & -1 & 0 & 3 & 0 & 1 \\ 1 & 0 & 1 & 0 & -1 & 0 & -1 & 0 & 0 \end{bmatrix}.$$

Again, $\boldsymbol{b}_3 = \frac{1}{\sqrt{2}}(\boldsymbol{b}_0 + i\boldsymbol{b}_1)$ results from superposing the initial states with a 90-degree phase shift on the latter. Taking the squared norms of its entries gives

$$\frac{1}{32}\begin{bmatrix} 0 & 0 & 2 & 0 & 2 & 0 & 10 & 0 & 2 \\ 2 & 0 & 10 & 0 & 2 & 0 & 2 & 0 & 0 \end{bmatrix}.$$

Summing the columns finally yields the desired distribution D of classical probabilities:

$$D = \left[\frac{1}{16}, 0, \frac{3}{8}, 0, \frac{1}{8}, 0, \frac{3}{8}, 0, \frac{1}{16}\right].$$

This is the surprise: the five nonzero numerators differ from $[\frac{1}{16}, \frac{1}{4}, \frac{3}{8}, \frac{1}{4}, \frac{1}{16}]$, which is the classical walk's binomial distribution. The quantum coin-flip distribution is flatter in the middle, with more weight dispersed to the edges.

As n increases, this phenomenon becomes more pronounced: the locations near the origin cumulatively have low probability, whereas most of the probability is on nodes at distance proportional to n. This phenomenon persists under various quantum coin matrices. The general reason is that the many paths under the classical "H-T" indexing that end close to the origin tend to cancel themselves out, whereas the fewer classical paths that travel do enough mutual reinforcement that the squared norms compensate for the overall count.

The effect is that, unlike the classical distribution, which for n steps has standard deviation proportional to $\sqrt{n}$, the quantum distributions have standard deviation proportional to n. Some implementations make them approach the uniform distribution on $[\frac{-n}{\sqrt{2}}, \frac{n}{\sqrt{2}}]$, whose standard deviation is $\sqrt{\frac{1}{6}}n > 0.4n$, which is a pretty big factor of n. Thus, the quantum traveler does a lot of boldly going where it hasn't been before.

14.7 Problems

14.1. Verify the probabilities obtained for the four-step walk on the graph with nodes labeled -4 to $+4$. What happens when this walk is started in the state $\frac{1}{\sqrt{2}}(\boldsymbol{a}_0 + \boldsymbol{a}_1)$, that is, without the phase shift on the "initial heads" part?

14.2. Work out the amplitudes and probabilities for the three-step walk with the $\boldsymbol{J}$ matrix in place of the Hadamard action on the coin space.

14.3. Can you devise a combinatorial rule for figuring the destination and amplitude of basic paths under the $\boldsymbol{J}$ matrix, in terms of the binary code of the path, analogous to the counting of HH substrings for the $\boldsymbol{H}$ matrix?

14.4. Can you use the combinatorial rule for the $\boldsymbol{H}$ matrix to prove that the amplitude of $(0, 0)$ for an n-step walk (n even) on the infinite path graph is bounded by $\frac{1}{\sqrt{2n}} + \epsilon$, for suitable $\epsilon > 0$?

14.8 Summary and Notes

The key to understanding a classical random walk in a graph is that it is defined locally. That is, for every node u and every neighbor v of u, define $\mathbf{A}(u,v)$ to be the probability of choosing to walk to v. Then $\mathbf{A}^2(u,w)$ is the probability of reaching node w in exactly *two* steps given that you started at u, and this carries over to any power: $\mathbf{A}^k(u,v)$ gives the probability of ending at v in exactly k steps, given that you started at u. Thus, random walks in graphs are just linear algebra.

The nice thing about quantum walks is that they too are just linear algebra, except the entries are complex numbers whose squared absolute values become the probabilities. The use of matrix multiplication and summing over paths is the same, except that what gets summed are possibly complex amplitudes rather than probabilities. The difference—maybe not so nice—is that these amplitudes can cancel, thus giving *zero* probability for certain movements from u to w that would be possible in the classical case. This enables piling higher probabilities on other movements in ways that cannot be directly emulated classically. But again the key is that the definition is local for each "node" (which is just a basis state) and gives you a matrix.

We have explained quantum coins with a "hidden-variables" mentality, and one must be careful when combining that with ideas of "local." However, this view has not tried to obscure nonlocal phenomena such as the way certain superpositions, for instance, the two coin states without the phase shift, can render certain far locations unreachable. We have really tried to emphasize the role of linear algebra and combinatorial elements such as the enlarged graph G' and the counting of HH substrings. It may seem strange to picture that nature tracks the parity of substring counts, but this is evidently the effect of what nature does. There are formalisms we haven't touched such as calculating in the Fourier-transformed space of the walk branches, which can avoid an exponential amount of work while at least giving good approximations.

There are some undirected graphs of small *girth*, that is, maximum distance between any pair of nodes, in which a quantum walk diffuses exponentially faster than a classical one. One such graph glues together two full binary trees of depth d at the leaves, giving girth $2d$ and $n = 3 \cdot 2^d - 2$ nodes overall. A classical random walk starting from one root quickly reaches the middle but thereafter has two ways to turn back for every one way forward toward the other tree's root, giving hitting time exponential in d although still $O(n)$. In the corresponding quantum walk, the turn-back options can be made to interfere enough that the situation essentially becomes a walk on the straight path, giving $O(d)$ diffusion as above. These and many other results about walks have recently been comprehensively surveyed by Venegas-Andraca (2012). Classical random walks on undirected graphs were "de-randomized" via the famous deterministic logarithmic space algorithm for graph connectivity of Reingold (2005).

Quantum walks have yielded improved and optimal algorithms for certain decision problems. Beyond that, they exemplify the idea that quantum computation is about creating rich quantum states, one that cannot be readily simulated by classical means, by which novel solutions can be obtained.

This chapter has followed some material in the survey paper by Kempe (2003), but with some more elementary examples and a bridge to the material of Santha (2008) and Magniez et al. (2011) incorporated in the next chapter. There is also a recent textbook by Portugal (2013) devoted entirely to quantum walks and search algorithms; it is comprehensive for this chapter and part of the next.

15 Quantum Walk Search Algorithms

The last chapter gave an elementary, self-contained treatment of quantum walks. The present chapter brings us to developments in quantum algorithms within the past 5 to 10 years. Our main purpose is to leverage the ideas of the last two chapters to explain a "meta-theorem" that underlies how quantum walks serve as an algorithmic toolkit for search problems.

15.1 Search in Big Graphs

We have seen that at least on the path graphs, a quantum walk does a good and fast job of spreading amplitude fairly evenly among the nodes, rather than lumping it near the origin as with a classical random walk. When the graph G is bushier and has shorter distances than the path graph, we can hope to accomplish such spreading in fewer steps. If we are looking for a node or nodes with special properties, then we can regard the evened-out amplitude as the springboard for a Grover search. If the graph's distances relative to its size are small, then we can even tolerate the size becoming super-polynomial—provided the structure remains regular enough that the action of a coin with d basic outcomes can be applied efficiently at any node of degree d.

For motivation we discuss the following problem, whose solution by Andris Ambainis is most credited for commanding attention to quantum walks.

Element distinctness: Given a function $f\colon [n] \to [n]$, test whether the elements $f(x)$ are all distinct, i.e., f is 1-to-1.

The best-known classical method is to sort the objects according to $f(x)$ and then traverse the sorted sequence to see whether any entry is repeated. If we consider evaluations $f(x)$ and comparisons to take unit time, then this takes time proportional to $n \log n$.

If we wish to apply Grover search, then we are searching for a *colliding pair* (x, y), $y \neq x$, such that $f(x) = f(y)$. We can implement a Grover oracle for this test easily enough, but the problem is that there are $\binom{n}{2}$ = order-n^2 pairs to consider. Thus, the square-root efficiency of Grover search will merely cancel the exponent, leaving $O(n)$ time, which is no real savings considering that the encodings of x, y and values of f really use $O(\log n)$ bits each.

The idea is to make a bigger target for the Grover search. Let $r > 2$ and consider subsets R of r-many elements. Call R a "hit" if f fails to be 1-to-1 on R. Testing this might seem to involve recursion, but we will first expend r

quantum steps to prepare a superposition of states that include the values $f(u_i)$ for every r-tuple $(u_1, \ldots, u_r) \in R$ in a way that the hit-check for every R is recorded. Thus, the preparation time for the walk is reckoned as proportional to r, and the hit-check needs no further evaluation of f.

However, now we have order-n^r many subsets, which seems to worsen the issue we had with Grover search on pairs. This is where quantum walks allow us to exploit three compensating factors:

1. The subsets have a greater density of "hits": any $r-2$ elements added to a colliding pair make a hit. Hence, the hit density is at least

$$\frac{\binom{n-2}{r-2}}{\binom{n}{r}} = \frac{r(r-1)}{n(n-1)} \approx \left(\frac{r}{n}\right)^2.$$

   In general, we write E for the reciprocal of this number.
2. Rather than make a Grover oracle that sits over the entire search space, we can make a quantum coin that needs to work only over the d neighbors of a given node.
3. If we make a degree-d graph that is bushy enough, then we can diffuse amplitude nearly uniformly over the whole graph in a relatively small number of steps.

The walk steps in the last point represent an additional time factor compared with a Grover search, but the other two points reduce the work in the iterations. This expounds the issues for search in big graphs. Now we are ready to outline the implementation.

In thinking about the element-distinctness example for motivation in what follows, note that the vertices of the graph G are not the individual elements x, y but rather the (unordered) r-tuples of such elements, corresponding to sets R. The adjacency relation of G in this case is between R and R' that share $r-1$ elements, so that R' is obtained by swapping one element for another. This defines the so-called **Johnson graph** $J_{n,r}$. Although all our examples are on Johnson graphs, the formalism in the next section applies more generally.

The final major point is that to encode an r-tuple as a binary string, we need $\Theta(r \log n)$ bits, and hence at least that many qubits. Because $r = r(n)$ will often be n^c for some constant c, this is a higher order than a compact encoding of $[n]$ as $\{0,1\}^\ell$ via $\ell \sim \log_2 n$ qubits would give us. This is why we call the graphs "big."

The main impact is that we will not be able to hide factors of r under O-tilde notation the way we can with factors of ℓ. This will entail distinguishing the following three chief cost measures:

1. Quantum **serial time**, which is identified with the quantum circuit size;
2. Quantum **parallel time**, which is the quantum circuit *depth* meaning the maximum number of basic gates involving any one qubit; and
3. Quantum **query complexity**, which is the number of evaluations of the oracle $\boldsymbol{U}_f$ for f.

Usually these costs follow this order from highest to least. Provided $f(u_i)$ is computable (classically) in time $|u_i|^{O(1)} = \ell^{O(1)}$ for any element u_i in a tuple, one can hope for either time to be O-tilde of the query complexity. The main issue is looking up a desired or randomly selected u_i, sometimes also a stored function value $f(u_i)$, in a tuple. If the tuple is sorted, then binary search will work in $O(\log r)$ stages and $O(\ell \log r) = \ell^{O(1)}$ depth (the literature also speaks of "random-access time"). However, the need to use gates on all r elements can cause an extra factor of r in the circuit size or serial time.

15.2 General Quantum Walk for Graph Search

The first idea in formulating quantum walks generically is that the coin space need not be coded as $\{1, \ldots, d\}$ but can use a separate copy of the node space. Then the expanded graph G' becomes the **edge graph** of G. We still reference nodes $X, Y, Z \ldots$ in our notation; the capital letters come from thinking of G as a big graph as above. The previous node X of a walk now at node Y is preserved as with the previous coin state in the current state (X, Y). Execution of a step to choose a next node Z is achieved by changing (X, Y) to (Y, Z). This can be done by treating the first coordinate as the "coin space" and replacing X by a random Z to make (Z, Y), then either permuting coordinates to make (Y, Z) explicitly or leaving (Z, Y) as-is and being sure to treat the other coordinate with Y as the coin space next.

The next two ideas come out of Grover search. As before, the goal is to concentrate amplitude on one or more of the "hit" nodes. This is the reverse of the notion of a walk starting from that node or nodes, which would diffuse out to uniform probability. Reversal, however, is "no problem" for quantum computation. Hence, we can start up in a stationary distribution of the classical walk on G, which by the first idea will extend naturally to the quantum walk because G' uses a copy of the nodes of G. For a d-regular graph, the uniform distribution is stationary.

Of course we cannot expect that a generic walk will run in reverse to every possible start node because that would not be invertible. The third idea is to combine the diffusion step with a Grover-type sign flip upon detecting that a node reached on the current edge is a hit. This will drive amplitude onto the hit nodes. Accordingly, we define a diagonal unitary operator $\boldsymbol{U_f}$ on our doubled-up graph space as a matrix with diagonal entries

$$\boldsymbol{U_f}[XY] = \begin{cases} -1 & \text{if } X \text{ or } Y \text{ is a hit,} \\ 1 & \text{otherwise.} \end{cases}$$

The game now becomes: how can we minimize effort while computing this $\boldsymbol{U_f}$, and how can we identify it with the Grover reflection-rotation scheme? We can define the *hit vector* $\boldsymbol{h}$ and the orthogonal *miss vector* $\boldsymbol{m}$ much as before, except now we have $\boldsymbol{h}(XY) = 1$ (suitably normalized) whenever X or Y is a hit, $\boldsymbol{m}(XY) = 1$ when neither is. The stationary distribution π of the walk on the paired-up graph space will give us a quantum start vector π analogous to $\boldsymbol{j}$ in chapter 13, again belonging to a two-dimensional space S_0. The following, however, causes tension between the two objectives:

> To minimize the cost of updates for the walk, we use functionally superposed states $s_Y = \sum_X |XY\rangle|g(XY)\rangle$ where $|g(XY)\rangle$ is "data" to facilitate the checking and possibly the updating steps. However, as explored in problem 13.9 in chapter 13, the values $g(XY)$ in the extended space S_g upset the geometry of two reflections producing a rotation toward $\boldsymbol{h}$ that we would enjoy in the simple XY space S_0 if g were absent (or constant).

A trick we could apply to solve the immediate problem is the "compute-uncompute" trick in section 6.3, as employed by the Grover approximate counting algorithm in section 13.5. In this case, we would first *uncompute* $|XY\rangle|g(XY)\rangle$ to $|XY\rangle|0\cdots0\rangle$, apply the reflections in S_0, and then *recompute* the data from scratch to set up the next walk step. However, this defeats the purpose of using the neighborhood data $g(XY)$ to make the walk more efficient because uncomputing and recomputing the neighborhood data items from scratch is more expensive than simply updating them.

The best solution known today works in a hybridization of S_g and S_0 that *approximates* the reflection action in S_0. The detailed proof that the approximation is close enough to make the rotation mechanism work is beyond our scope (see end notes for references). Fortunately, the theorem that this proves has a nice "toolkit" statement so that applications can spare these details and concentrate on the cost parameters—which can accommodate any of our three complexity measures. Here is our first statement of the five parameters:

- S: the **setup cost** for the functional superpositions giving "data" $D = g(XY)$.
- U: the **update cost** for a walk step in the graph, which when repeated will approximate a reflection about π. By intent it equals the cost for a step in the underlying classical walk, although sometimes this involves blurring a distinction between $O(1)$ and $O(\log n)$ or $O(\log d)$ for the update.
- C: the **checking cost** for (the reflection effecting) each Grover query $\boldsymbol{U_f}$.
- E: the **reciprocal density** of hits after the initial setup, i.e., $\mathsf{E} = N/k$.
- D: the reciprocal of the **eigenvalue gap** Δ defined below, which governs the spread of the walk.

The next two sections give all the details on how the walk is implemented so that the algorithms governed by these cost parameters can **succeed**.

15.3 Specifying the Generic Walk

We first implement the generic walk in the form of reflections. Let $p_{X,Y}$ give the probability of going from X to Y in the standard classical walk on G and $p^*_{Y,X}$ the probability in the reverse walk. For a stationary distribution π of the forward walk on $V(G)$, $p^*_{Y,X}$ obeys the equation $\pi_Y p^*_{Y,X} = \pi_X p_{X,Y}$. The walk is **reversible** if in addition $p^*_{X,Y} = p_{X,Y}$. On a regular graph π is uniform distribution, so a reversible walk gives $p_{Y,X} = p^*_{Y,X} = p_{X,Y}$ and hence is also **symmetric**. But we can define the following "right" and "left" **reflection operators** even for a general walk:

$$
\begin{aligned}
\boldsymbol{V_R}[XY,XZ] &= 2\sqrt{p_{X,Y}p_{X,Z}} - e(Y,Z) \\
\boldsymbol{V_L}[XZ,YZ] &= 2\sqrt{p^*_{Z,X}p^*_{Y,Z}} - e(X,Y).
\end{aligned}
$$

Here $e(Y,Z)$ is a discrete version of the Dirac "delta" function giving 1 if $Y = Z$ and 0 otherwise. If $W \neq X$, then $\boldsymbol{V_R}[XY,WZ] = 0$, and similarly $\boldsymbol{V_L}[XZ,YW] = 0$.

We can define these operators more primitively by coding the walk directly. Let Y_0 be some basis state in the coin space—usually it is taken to be the node coded by the all-zero string, but this is not necessary. Take $\boldsymbol{P_R}$ to be any unitary operator such that for all X and Y,

$$\boldsymbol{P_R}[XY, XY_0] = \sqrt{p_{X,Y}}.$$

For $Z \neq Y_0$, $\boldsymbol{P_R}[XY, XZ]$ may be arbitrary, but $\boldsymbol{P_R}[XY, WZ] = 0$ whenever $W \neq X$. Among concrete possibilities for $\boldsymbol{P_R}$, we could let Y_0 be the state coded by the all-1 string instead and control on it so that the action on $Z \neq Y_0$ is the identity, or we could define $\boldsymbol{P_R}[XY, XZ] = \sqrt{p_{X,Y \oplus Y_0 \oplus Z}}$. It turns out not to matter. Define the projector and reflection about Y_0 by

$$\begin{aligned}
\boldsymbol{P_0}[XY, WZ] &= \begin{cases} 1 & \text{if } W = X \wedge Y = Z = Y_0 \\ 0 & \text{otherwise;} \end{cases} \\
\boldsymbol{J_0}[XY, WZ] &= 2\boldsymbol{P_0}[XY, WZ] - e(XY, WZ) \\
&= \begin{cases} 1 & \text{if } X = W \wedge Y = Z = Y_0, \\ -1 & \text{if } X = W \wedge Y = Z \neq Y_0, \\ 0 & \text{otherwise.} \end{cases}
\end{aligned}$$

Our two unitary reflections hence enjoy the following equalities:

$$\begin{aligned}
\boldsymbol{V_R} &= \boldsymbol{P_R J_0 P_R^t} \\
\boldsymbol{V_L} &= \boldsymbol{P_L J_0 P_L^t}.
\end{aligned}$$

Now the following blends a Grover search with the quantum walk:

DEFINITION 15.1 The **generic quantum walk** derived from the classical walk $P = (p_{X,Y})$ on the graph G is the walk on $V(G) \otimes V(G)$ defined by iterating the step operation

$$\boldsymbol{W_P} = \boldsymbol{V_L V_R},$$

and the **generic search algorithm** iterates alternations of $\boldsymbol{U_f}$ and $\boldsymbol{W_P}$, or more generally, $\boldsymbol{U_f}$ followed by a routine involving one or more walk steps.

As a footnote, there are allowable variations in $\boldsymbol{U_f}$ as well as the walk steps. Provided G is not bipartite, it is OK to omit a check for Y being a hit in the definition of $\boldsymbol{U_f}$. That is, if we define

$$\begin{aligned}
\boldsymbol{U_L}[XY] &= -1 \text{ if } X \text{ is a hit,} \quad \boldsymbol{U_L}[XY] = 1 \text{ otherwise;} \\
\boldsymbol{U_R}[XY] &= -1 \text{ if } Y \text{ is a hit,} \quad \boldsymbol{U_R}[XY] = 1 \text{ otherwise,}
\end{aligned}$$

then it suffices to iterate $\boldsymbol{W_P U_L}$ without checking whether we found a hit in the Y coordinate. Perhaps more elegantly, we can iterate $\boldsymbol{V_L U_L V_R U_R}$. We do not know whether the concrete effects on implementations have been all worked out in the recent literature, likewise with particular choices for $\boldsymbol{P_R}$ and the analogous $\boldsymbol{P_L}$, but asymptotically they give equivalent results.

15.4 Adding the Data

The data can be specified separately for the nodes X, Y in a graph edge, and we find it intuitive to separate $g(XY)$ as D_X, D_Y for indexing. To encode the data in extra indices, we can re-create all the above encoding of operators with XD_X in place of X. Everything goes through as before—technically because we will have $p_{XD_X,YD_Y} = 0$ whenever YD_Y does not match the data update when going from node X with D_X to node Y.

These considerations also factor into the initial state of the walk. Let X_0 be the node coded by the all-zero index, Y_0 the same but on the right-hand side of indexing XY, and D_0 the data associated to Y_0. One possibility for the initial state $\boldsymbol{a}_0$ of the walk is defined by

$$\boldsymbol{a}_0(XY_0) = \sqrt{\pi_X},$$

and $\boldsymbol{a}_0(XY) = 0$ for $Y \neq Y_0$. This can be interpreted either as starting at the all-zero node with the coin in a stationary-superposed state or having the coin initialized to "zero" with the walk initially in the classical stationary superposed state. On a regular N-node graph G, we always have $\pi_X = 1/N$. Now if we throw in the data, this technically means initializing the state

$$\boldsymbol{a}_0(XD_X Y_0 D_0) = \sqrt{\frac{1}{N}},$$

with $\boldsymbol{a}(XDYD') = 0$ whenever $D \neq D_X$, $Y \neq Y_0$, or $D' \neq D_0$. Because getting uniform superpositions is relatively easy, and the adjacency relation of G is usually simple, the difficulty in preparing $\boldsymbol{a}_0$ actually resides mainly in determining the associated data. The same applies to the update cost—the time taken to implement the concrete $\boldsymbol{V_R}$ and $\boldsymbol{V_L}$ via extended versions of $\boldsymbol{P_R}$ and $\boldsymbol{P_L}$ is mainly for getting the data associated with the node Y traversed from X.

Finally, in the concrete version of the "flip" $\boldsymbol{U_f}$, there is the cost of checking the associated data to see whether the current node is a "hit." For these steps and checks to be done in superposition, they must be coded for accomplishment by linear transformations. All of this is done by jockeying indices.

Our point is not so much to argue that our notation is more intuitive or less cumbersome than standard notation—such as $\sum_X |XY_0\rangle\langle XY_0|$ for the projector onto $\mathbb{C}^{V(G)} \otimes |Y_0\rangle$, or $|X\rangle|D_X\rangle$ to carry along the associated data—but rather to explore lower level details and suggest possible alternate encodings.

15.5 Toolkit Theorem for Quantum Walk Search

The last factor for efficient quantum walks is that the underlying graph G be sufficiently "bushy" relative to its size N. The formal notion is that G be an **expander**, meaning that there is an appreciably large value $h(G)$ such that for all sets T of at most $N/2$ nodes, there are at least $h(G)|T|$-many edges going to nodes outside T. This implies that there are at least $h(G)|T|/d$ different nodes outside T that can be reached in one step, but it is separately significant to have many edges that can diffuse amplitude into these neighboring nodes. An important lower bound on $h(G)$ is provided by the difference between the largest eigenvalue and the second largest absolute value of the adjacency matrix of G, which is called the *eigenvalue gap* and denoted by $\gamma(G)$.[1] The bound is

$$\frac{1}{2}\gamma(G) \leq h(G).$$

For a d-regular graph, we finally define

$$\mathsf{D} = \frac{d}{\gamma(G)}.$$

This is also the reciprocal of the eigenvalue gap δ of the stochastic matrix of the underlying standard classical walk on G. Note that as with $\mathsf{E} = N/k$, we are suppressing the dependence on the original problem-size parameters n and d with this notation. The point of doing this, and likewise with the setup cost $\mathsf{S} = \mathsf{S}(n)$, the update cost $\mathsf{U} = \mathsf{U}(n)$, and the solution-checking cost $\mathsf{C} = \mathsf{C}(n)$, is that the total cost of many quantum walk search algorithms can be expressed entirely in these five terms. The following "toolkit" main theorem applies to any cost measure that satisfies some minimal assumptions, such as being additive when routines are sequenced, and in particular applies to measures of time, quantum circuit size, quantum circuit "depth" (i.e., the maximum number of gates on any qubit, which serves as an idea of parallel time), and the count of superposed queries to the search function f.

1 More common is $\Delta(G)$, but our sources use $\Delta(P)$ for the *phase gap* of the quantum walk as discussed below.

THEOREM 15.2 For any search problem on a uniform family of undirected graphs G_n whose adjacency matrices yield walks with diffusion parameters D, one can design a quantum walk with setup, update, and checking phases, such that if their respective costs are bounded by $\mathsf{S}, \mathsf{U}, \mathsf{C}$, respectively, and if the initial reciprocal hit density is E, then the overall cost to achieve correctness probability at least $3/4$ is bounded above by a constant times

$$\mathsf{S} + \sqrt{\mathsf{E}}(\mathsf{U}\sqrt{\mathsf{D}} + \mathsf{C}). \tag{15.1}$$

We sketch the proof as an algorithm, mindful of detail dropped in one step, and of D and the other parameters really being functions of n.

15.5.1 The Generic Algorithm

As in chapter 13, we first state the algorithm in the case that the number k of hits—and hence E—is known. The revision for the unknown-k case is essentially the same as in section 13.4, so we omit it. The constants c and c' will come from details of the approximation step that we skip, but it is enough to know that they are reasonable constants.

1. The start vector $\boldsymbol{a}_0$ is the superposition π derived from the stationary distribution of the underlying classical random walk, augmented with data into a functional superposition. That is,

$$\boldsymbol{a}_0 = \sum_{X,Y} \sqrt{\pi_{X,Y}} |X\rangle |D_X\rangle |Y\rangle |D_Y\rangle$$

   This is further tensored with $c'\sqrt{\mathsf{E}}\log(\mathsf{D})$ ancilla qubits set to 0, which are used to run the approximation routine and offset deviations from "true" reflections that are caused by the data qubits.

2. Repeat $c\sqrt{\mathsf{E}}$ times:

   2.1 Apply the Grover reflection about the "miss" vector in the form of a sign flip, which for reasons discussed in section 6.5 is not affected by the extra "data" and ancilla coordinates.

   2.2 Apply a recursive procedure, whose details from Magniez et al. (2011) we elide, whose major part and cost involves $c'\sqrt{\mathsf{D}}$ steps of the quantum walk, which together approximate a reflection about π.

3. Measure the final state $\boldsymbol{a}$, giving a string $x \in \{0,1\}^n$.

4. If x is a solution, then stop. Otherwise repeat the entire process.

15.5.2 The Generic Analysis

Proof sketch for theorem 15.2. By the definition of the setup and checking costs, step 1 costs S and each iteration of step 2.1 costs C. The latter holds by definition even when step 2.1 is a complicated routine that involves branching or recursion, as in examples to come.

Although we have elided details, the point in step 2.2 is that the efficacy of a *quantum* walk P is that its *phase gap* $\Delta(P)$ is proportional (at least) to the *square root* of the eigenvalue gap δ of the underlying classical walk. This channels the observation about the quadratic advantage in spreading made at the end of chapter 14. In our form using the reciprocal notation D, this comes out as cost bounded by $c'\sqrt{\mathsf{D}}$. Originally the "c'" involved a factor of $\log(\mathsf{E})$, but recursive use of *phase estimation* by Magniez et al. (2011) avoids it. Hence, the total time for the algorithm is of order

$$\mathsf{S} + \sqrt{\mathsf{E}}(\mathsf{U}\sqrt{\mathsf{D}} + \mathsf{C}).$$

Modulo the (substantial) omitted details, this proves theorem 15.2. □

The way to *apply* this theorem is to find appropriate graphs on which to model the search problem at different problem sizes n, get the $\mathsf{D}(n)$ and $\mathsf{E}(n)$ values from the graphs, and design additional features associated to the nodes (if needed) to balance out $\mathsf{S}(n)$, $\mathsf{U}(n)$, and $\mathsf{C}(n)$. Before showing how this theorem plays out in some new examples, we pause to review Grover search in this formalism.

15.6 Grover Search as Generic Walk

Here G_N is the complete graph on N vertices, $N = 2^n$. The graph is implicit, with vertices $x \in \{0,1\}^n$ being the only representation. Because G_N has degree $N-1$ and the next highest eigenvalue is known to be 1, the gap is $N-2$. Hence, $\mathsf{D}(n) = (N-1)/(N-2) \approx 1$. Suppose at least k of the N nodes are hits. Then $\mathsf{E}(n) \leq N/k$.

The setup takes one stage of n parallel Hadamard gates, which can be reckoned as n if counting basic gates as a measure of sequential time, $O(1)$ if counting circuit depth as a notion of parallel time, or 0 if counting queries to f, i.e., applications of $\boldsymbol{U}_f$. The checking cost is one such query but may be best regarded as $n^{O(1)}$ for both parallel and sequential time. The update cost can be reckoned as 0 if counting queries, 1 if counting reflections, or $O(n)$ if counting the number of basic gates for the reflection about j. Using sequential time as the cost measure, we have:

- $\mathsf{S} = O(n)$;
- $\mathsf{U} = O(n)$;
- $\mathsf{C} = n^{O(1)}$;
- $\mathsf{E} = \frac{2^n}{k}$;
- $\mathsf{D} = O(1)$;

$$T \;=\; \mathsf{S} + \sqrt{\mathsf{E}}(\mathsf{U}\sqrt{\mathsf{D}} + \mathsf{C}) \;=\; O(n) + \sqrt{\frac{2^n}{k}}(O(n) + n^{O(1)}) \;=\; \tilde{O}(\sqrt{\frac{N}{k}}).$$

This accords with theorem 13.1. In particular for $k = 1$, the time is $\tilde{O}(2^{n/2})$. Note that the quantum walk architecture for controlling the search preserves the guarantee of faster time if the number k of hits is large. If we are counting queries, we can simplify by regarding $f(x)$ as data associated to the node x, giving $\mathsf{U} = 1$, $\mathsf{C} = 0$, and $T = O(\sqrt{N/k})$ queries.

15.7 Element Distinctness

Recall the problem is: given a function $f\colon [n] \rightarrow [n]$, test whether the elements $f(x)$ are all distinct, i.e., f is 1-to-1. Also recall from section 15.1 our intent to amplify by taking r-tuples of elements as nodes of a graph, so going beyond $\ell \sim \log_2 n$ qubits.

Several kinds of graphs work, but the Johnson graphs $J_{n,r}$ are the original and most popular choice. Recall that $J_{n,r}$ has a node for each $R \subset [n]$ of size r, and edges connecting R, R' when they have $r-1$ elements in common; note that the complete graph equals $J_{n,1}$. The degree of $J_{n,r}$ is $r(n-r)$ because every edge involves deleting one element u from R and swapping in one element v not in R. The second eigenvalue is $r(n-r) - n$, so the gap is n provided $r \geq 2$. Thus, $\mathsf{D}(n) = \frac{r(n-r)}{n} = r - r^2/n < r$. From the above hit density we have $\mathsf{E}(n) = (n/r)^2$. The goal is to choose r as a function of n to balance and minimize the total cost.

The quantum algorithm needs to initialize not only a uniform superposition over nodes $R = (u_1, \dots, u_r)$ but also the values $f(u_1), \dots, f(u_r)$ for its elements—and to sort the latter locally. This requires r linearly superposed queries to f and $\tilde{O}(r)$ time overall assuming $f(u_i)$ is in time $|u_i|^{O(1)}$. The update needs to remove the value $f(u_i)$ and add a value $f(v)$ when u_i is swapped out for v in the walk. Here is mainly where the cost measures diverge. The query cost is just 2, and a binary search to maintain sortedness of the tuple and do the update can work in $O(\log r)$ stages and $O(\ell \log r)$ random-access/parallel time overall. The need for a circuit, however, to touch all r elements in a tuple takes serial time out of our savings picture. Thus, we keep tabs on the query and parallel time complexities, and we have:

- $\mathsf{S} = \tilde{O}(r)$;
- $\mathsf{U} = \ell^{O(1)}$, with two queries;
- $\mathsf{C} = \ell^{O(1)}$, with zero queries because results are kept in the data;
- $\mathsf{E} = \left(\frac{n}{r}\right)^2$;
- $\mathsf{D} \leq r$;

$$T = \mathsf{S} + \sqrt{\mathsf{E}}(\mathsf{U}\sqrt{\mathsf{D}} + \mathsf{C}) = \tilde{O}(r) + \frac{n}{r}\tilde{O}(\sqrt{r}) = \tilde{O}(r + \frac{n}{\sqrt{r}}).$$

This is balanced with $r = n^{2/3}$, giving parallel time $\tilde{O}(n^{2/3})$ and query complexity a clean $O(n^{2/3})$. It is known that $\Omega(n^{2/3})$ queries are necessary, so this bound is asymptotically tight ("up to tilde" on parallel time).

15.8 Subgraph Triangle Incidence

Given an m-node subgraph H of an n-node graph G, does H have an edge of a triangle in G? Given a fixed edge (u, v) in H, we could do a $\sqrt{n}$-time Grover search for w such that (u, w) and (v, w) are also edges. But iterating this through possibly order-m^2 edges in H is clearly prohibitive. We will apply the savings from our major example of element distinctness and, hence, focus on query complexity and parallel time.

For each w, we instead do a quantum walk to find (u, v). We take $r = m^{2/3}$ and define a subset R of the nodes of H to be a hit if it contains a suitable edge (u, v). Using a walk on the Johnson graph $J_{m,r}$ plus data consisting of whether $(u, v), (u, w), (v, w)$ are all edges, all the parameters are the same as for element distinctness, so the parallel time is $\tilde{O}(m^{2/3})$.

This in turn becomes the checking time for the Grover search. Using the formula again, the overall parallel time is $\tilde{O}(n^{1/2}m^{2/3})$. We can amplify both the check and the Grover success probability to be at least $7/8$, so as to yield $3/4$ on the whole.

15.9 Finding a Triangle

Now we call an m-node subset R of $V(G)$ a hit if the induced subgraph H includes an edge of a triangle in G. The E and D will hence be the same as for element distinctness with "m" replacing "r." We may use the previous item to obtain checking cost $\mathsf{C}(n) = \tilde{O}(n^{1/2}m^{2/3})$. The one thing that is different is that the setup and update costs are higher. The setup needs to encode the entire adjacency matrix of H, and when the update swaps in a vertex v' and swaps out a v, it needs to update the $m-1$ adjacencies of v' while erasing those of v. Thus, we have setup cost $\tilde{O}(m^2)$ and update cost $\tilde{O}(m)$, giving:

- $\mathsf{S} = \tilde{O}(m^2)$;
- $\mathsf{U} = \tilde{O}(m)$;
- $\mathsf{C} = \tilde{O}(n^{1/2}m^{2/3})$;
- $\mathsf{E} = \left(\frac{n}{m}\right)^2$;
- $\mathsf{D} \leq m$;

$$T \;=\; \mathsf{S} + \sqrt{\mathsf{E}}(\mathsf{U}\sqrt{\mathsf{D}} + \mathsf{C}) \;=\; \tilde{O}(m^2) + \frac{n}{m}(\tilde{O}(m\sqrt{m} + n^{1/2}m^{2/3})).$$

This time, the setup cost drops out of the equation—the balancing is between the update and checking cost and is achieved when $m^{3/2} = n^{1/2}m^{2/3}$, that is, when $m^{5/6} = n^{1/2}$, so $m = n^{3/5}$. This results in the overall parallel time

$$\tilde{O}(n^{6/5} + n^{2/5}n^{9/10}) = \tilde{O}(n^{13/10}).$$

The query complexity is just $O(n^{13/10})$. References cited in the end notes have since improved the query complexity to $O(n^{9/7}) = O(n^{1.2857\ldots})$ by other methods, although the effects on various reckonings of time are less clear.

The best known classical algorithm for finding a triangle takes the adjacency matrix $\boldsymbol{A}$, squares it, and hunts for i,j such that $\boldsymbol{A}^2[i,j] \cdot \boldsymbol{A}[j,i] > 0$. It hence runs in time $\tilde{O}(n^{\omega})$, where the exponent ω of matrix multiplication is known to be at most 2.372. Is there any square-root relation whereby $\omega/2$ might become the best approachable exponent for the query complexity or (parallel) quantum time? Currently that would be a target of $O(n^{1.185})$. No quantum lower bound higher than linear is known for triangle detection; whereas no lower bound higher than two is known for ω. See the end notes for further references.

15.10 Evaluating Formulas and Playing Chess

Of course we must end by playing chess. Indeed, the original Grover search problem could be stylized as that of finding a winning move in a chess position, when the fact that it is winning can be verified instantly once you see it, but you are a poor enough player that even a checkmate is hard to find. If there are L possible moves—and if we assume unit time per move even on a board with n squares where L depends on n—then a Grover search says we can find the move in $\tilde{O}(\sqrt{L})$ expected time.

When the win is not instantly verifiable, however, we are in a much harder situation than a Grover search. Suppose we know in advance a bound m on the number of turns the game will last, counting moves by both players. Keeping L as the bound on the number of legal moves in any position, and waving away the event that different sequences of moves can lead to the same position, we have an L-way branching game tree of size about $N = L^m$. The classical time bound for exhaustive search to determine whether there is a winning strategy, one that is able to answer all possible moves by the opponent, is $T = O(L^m)$. Can a quantum algorithm improve this analogously to $O(\sqrt{T}) = O(L^{m/2})$?

We would like to use (15.1) to set up the following simple-minded recursion for the time $T = T(n,m)$ to traverse the game tree and find a winning move if one exists, reporting "no" if not. We may regard n as the maximum of the total board size and $\log N$. We pose the question, when is the following valid?

- $\mathsf{S} = O(n)$, immaterial as with a basic Grover search;
- $\mathsf{U} = O(1)$, because we merely play a move;
- $\mathsf{C} = 1 + T(n,m-1)$ by the desired recursion;
- $\mathsf{E} = L$ as for a Grover search in a space of size L;
- $\mathsf{D} = O(1)$;

$$\begin{aligned} T(n,m) &= \mathsf{S} + \sqrt{\mathsf{E}}(\mathsf{U}\sqrt{\mathsf{D}} + \mathsf{C}) = O(n + \sqrt{L}(1 + T(n,m-1))) \\ &\approx O(\sqrt{L} \cdot T(n,m-1). \end{aligned}$$

If we ignore log factors or just count queries to some table of checkmates, then the solution would indeed become $O(L^{m/2}) = O(\sqrt{N})$. The validity of this, however, raises issues that lead to recent research on quantum walks.

The first issue is amplification of the success probability, which this chapter has kept in the background. Even thinking in terms of a simple classical recursion, if the recursive call and the current-level Grover search are both tuned for success probability $3/4$, then the level falls to only a $9/16$ guarantee, which fails to maintain the recursion invariant. In Grover's quantum search, however, a constant error rate on the *miss* vector can have a huge effect. Without further tricks, it takes amplification of order $1 - \left(\frac{1}{4}\right)^m$ in the base step to contain this problem, and the solution originally obtained by Buhrman et al. (1998) gave time of order $L^{m/2}n^{m-1}$. Note that because the exponent m is not fixed, the n^m is not simply an O-tilde factor, and in chess (under an analogue of the "fifty-move rule"), m can be proportional to the total board size. Speedup needs $n^m \ll L^{m/2}$, and this is given only when $L > n^2$. These results also apply to evaluating Boolean formulas that alternate L-way OR and AND levels, as in Buhrman et al. (1998).

Second and more important, when L is small or even *constant*, it means little to say that each level's Grover-style walk is giving time $\sqrt{L}$. Trees of NAND gates cannot be grouped as trees of OR and AND and XOR can, so they are stuck with $L = 2$. For a long time, no quantum algorithm for evaluating a full depth-d tree of binary NAND gates was known to beat the $2^{0.753m}$ time achievable classically. However, first in an unconventional quantum model and then in our standard one, the $O(2^{m/2})$ target was approached and then achieved exactly in the count of queries (Ambainis et al., 2010, Reichardt, 2011b) together with getting the overall time down to $\tilde{O}(\sqrt{N})$ (Reichardt, 2011a). These improvements have come from further clever use of quantum phase estimation and deeper relationships to linear algebra (Reichardt and Špalek, 2008). Thus, for evaluating logic formulas, playing games of strategy, and various related problems, quantum algorithms can deliver the same near-quadratic speedup that we first saw for Grover's search algorithm.

15.11 Problems

15.1. Show by direct calculation that the matrix $\boldsymbol{V_R}$ is unitary.

15.2. Work out the analogous decomposition of the matrix $\boldsymbol{V_L}$ using an arbitrary basis state X_0 of the "left" node space. What if X_0 is not a basis state?

15.3. In the algorithm for element distinctness, suppose we use recursion to check whether a set of r nodes has two nondistinct function values and thus constitutes a "hit." Do you get the same $O(n^{2/3})$ running time or something less?

15.4. Sketch how to implement the element-distinctness algorithm on the *Hamming graphs* $H_{n,r}$, whose vertices are r-tuples from $[n]$ and whose edges connect r-tuples that differ in one place. Are all of $\mathsf{S}(n)$, $\mathsf{U}(n)$, $\mathsf{C}(n)$, $\mathsf{E}(n)$, and $\mathsf{D}(n)$ asymptotically the same as before?

15.5. Sketch a quantum search algorithm that given three $n \times n$ matrices $\boldsymbol{A}, \boldsymbol{B}, \boldsymbol{C}$ finds i, j such that $\sum_k \boldsymbol{A}[i,k]\boldsymbol{B}[k,j] \neq \boldsymbol{C}[i,j]$ if such a pair exists or else outputs "accept." You may consider arithmetic operations to be unit time and need only count steps that query matrix entries. The goal is cost $O(n^{5/3})$.

15.6. Suppose you can make superposed black-box queries to a binary operation $\circ$ on $[n]$ whose values lie in $[k]$. Sketch a quantum algorithm to find $a, b, c \in [n]$ such that $(a \circ b) \circ c$ is *not* equal to $a \circ (b \circ c)$ if any such "nonassociative triple" exists. If $k = O(1)$, then what is the running time of the algorithm? Note that the data associated to a Johnson-graph node $(u_1, \ldots, u_r)$ can preserve values of $\circ$ on arguments in $[k]$ as well as the u_i.

15.12 Summary and Notes

This chapter has followed the lead of Magniez et al. (2007) and Magniez et al. (2011), building on Szegedy (2004). We have chosen selectively from the survey by Santha (2008). We could have used the quantum amplitude amplification framework of Brassard et al. (1998) and Brassard et al. (2000) as a bridge from section 13.5 to section 15.6, where it would give cost $O(\sqrt{\mathsf{E}}(S + C))$, but we chose to keep both sections simpler and self-contained.

The last two problems also come from Santha (2008), while the problem about element distinctness on the Hamming graphs is based on Childs and Kothari (2012) and Childs (2013). The seminal paper on element distinctness appeared in full form as Ambainis (2007). The query complexity of triangle finding has since been improved from the $O(n^{13/10})$ in section 15.9 to $O(n^{9/7})$ by Lee et al. (2013), while connections to the exponent of matrix multiplication are explored further by Williams and Williams (2010). The advances for (playing chess and) evaluating NAND trees and other logical formulas include the work of Farhi et al. (2008), Childs et al. (2009), Reichardt and Špalek (2008), and Ambainis et al. (2010). See also the often-updated notes of Childs (2013).

Links to further lecture notes and surveys and texts, including Kitaev et al. (2002), Kaye et al. (2007), Childs and van Dam (2008), and Mosca (2009), may be found at the "Quantum Algorithms Zoo," which is maintained by Stephen Jordan (http://math.nist.gov/quantum/zoo/).

16 Quantum Computation and BQP

We have presented Shor's algorithm without giving a full statement of the theorem it proves. The theorem reads, "Factoring is in BQP." In words, this means that factoring has a feasible algorithm with bounded error. The letters BQP stand for **B**ounded-Error **Q**uantum **P**olynomial Time. This is the central complexity class in quantum complexity theory. This chapter defines it formally, shows that several other possible definitions are equivalent, and shows its relationship to longer-studied classes in "classical" complexity theory.

16.1 The Class BQP

We have already discussed the error probability of a quantum algorithm and how one can *amplify* the success probability. Saying "bounded error" entails formalizing conditions under which one can *amplify* the success probability of an algorithm. We define BQP for functions as well as decision problems. The **characteristic function** χ_L of a language L is defined by $\chi_L(x) = 1$ if $x \in L$ and $\chi_L(x) = 0$ otherwise.

DEFINITION 16.1 A function $f\colon \{0,1\}^* \to \{0,1\}^*$ belongs to BQP if there are a polynomial p, a function g computable in classical $p(n)$ time, and a quantum algorithm $\boldsymbol{A}$ such that for all n and inputs $x \in \{0,1\}^n$, and for some $r \leq p(n)$, $\boldsymbol{A}$ applied to the initial state $\boldsymbol{e}_{x0^r}$ yields within $p(n)$ basic quantum operations a quantum state $\boldsymbol{b}$ such that

$$\Pr[\text{measuring } \boldsymbol{b} \text{ yields } z \text{ such that } g(z) = f(x)] \geq \frac{3}{4}. \tag{16.1}$$

A language belongs to BQP if its characteristic function does.

The $3/4$ is arbitrary; it can be replaced by $1/2 + \epsilon$ for any fixed $\epsilon > 0$, and in some contexts where the relation $f(x) = y$ is already known to be in BQP, it can be lower. We have differed from standard sources in making the classical post-processing explicit. This accords better with our presentation of Shor's algorithm and makes the amplification of the success probability transparent.

To move from the general notion of a quantum *algorithm* to specific models of quantum *circuits*, however, we have to address the gates, control the success probability from measurements, and wean off the classical parts. The following theorem makes a statement doing so and, hence, removes our need to specify the underlying quantum machine or circuit model any further. A collection $[C_n]$ of circuits is **uniform** if the mapping from n to a description of C_n is computable in classical polynomial time.

THEOREM 16.2 For any set S of quantum gates that includes the Hadamard gate and adds either (i) the Toffoli gate, (ii) the controlled-phase gate ***CS***, or (iii) ***CNOT*** and the ***T***-gate, every function f in BQP is computable by uniform circuits $[C_n]$ of polynomially many gates in S, with "$\frac{3}{4}$" replaced by $1 - \epsilon(n)$ provided $\epsilon(n) \geq \exp(-n^k)$ for some k and all n, and with a single measurement without post-processing in which the value appears in the first $|f(x)|$-many qubit lines.

We can sketch much of the proof, although full details inevitably depend on whatever model one chose to specify quantum algorithms ***A*** to begin with. Various models were used until quantum circuits gained ascendancy. We could have obviated the missing details by using circuits of (i), (ii), and/or (iii) gates (and nothing else) as our model for quantum algorithms to begin with, but doing so would have cramped both history and this text's style.

Proof. Regarding the equivalence of the three gate sets, the details have been worked out in the exercises of chapters 6 and 7. That each of them is **universal**, meaning capable of close enough approximation of quantum circuits C using any other finite set of basic gates to preserve (16.1), also follows from ideas in these exercises as summarized in the end notes to chapter 6. Moreover, the **Solovay-Kitaev theorem** gives an algorithm to produce the new circuit efficiently while multiplying the size s of C by less than a constant times $(\log s)^4$. This algorithm applied for the gate sets (ii) or (iii) yields full approximate simulations of operators on complex Hilbert spaces, while for (i) using real spaces it just preserves the measurements needed for (16.1).

The exercises have also worked out how to decompose the quantum Fourier transform as a composition of $O(n^2)$-many one- and two-qubit gates. The representation obtained in problem 6.13 does not use a finite set of gates because the twists $\boldsymbol{T}_\alpha$ are used with the angles $\alpha = \pi/2^{n-1}$ being exponentially fine. However, the Solovay-Kitaev *process* also applies to these gates, and the same efficiency giving the $(\log s)^{O(1)}$ overhead enables it to achieve approximation as an operator on $\mathbb{C}^N$ using only $(\log N)^{O(1)} = n^{O(1)}$ gates from sets (ii) or (iii). This gives more than required because BQP need only satisfy (16.1) in the measurements, so the issues with complex angles get flattened out when everything is done in the real Hilbert space $\mathbb{R}^{2N}$.

It remains to discuss the amplification to error at most $1/2^{n^k}$. This comes from the ability to clone the basis state given as input into order-n^k many copies, run the quantum gates in parallel on the copies, measure to get a vector of outputs z_i, and take the majority vote of the final outputs $g(z_i)$ to yield

$f(x)$. The last and trickiest fact is that this probability can be amplified without relying on classical majority vote in a post-processing step. Instead, the idea of deferred measurement is employed to do repeated trials and accumulate results within the circuit. Because the majority vote is in classical polynomial time, theorem 5.4 is implicitly used to bring the post-processing with majority vote within the quantum circuit as well. Hence, a single measurement ultimately suffices to yield $f(x)$ with the amplified success probability. □

In the case of languages L, the conditions for $L \in \mathsf{BQP}$, together with the achievable amplification, look like this: For any input $x \in \{0,1\}^n$ together with $r = n^{O(1)}$ ancilla qubits, write $\boldsymbol{a}_x$ as short for $\boldsymbol{e}_{x0^r}$ and write $\boldsymbol{U}$ for the $2^{n+r} \times 2^{n+r}$ dimensional unitary transformation the circuit C computes. By the compute-uncompute trick in section 6.3, we can arrange for acceptance to yield $\boldsymbol{a}_x$ again as output. Then C being a BQP-circuit for L (with amplification) is equivalent to the conditions that for all $x \in \{0,1\}^n$:

$$\begin{aligned} x \in L &\implies |\boldsymbol{a}_x \cdot \boldsymbol{U}\boldsymbol{a}_x|^2 \geq 1 - \epsilon(n), \\ x \notin L &\implies |\boldsymbol{a}_x \cdot \boldsymbol{U}\boldsymbol{a}_x|^2 \leq \epsilon(n). \end{aligned}$$

This form facilitates comparing BQP with other complexity classes, which are most commonly defined in terms of languages. We can always associate a language L_f to a function f so that $f(x)$ can be computed efficiently via a subroutine for whether strings combining x with incrementally built binary strings w belong to L_f. We did this with the factoring problem in the first part of chapter 4. Thus, functions and languages are usually considered notionally equivalent in complexity theory.

16.2 Equations, Solutions, and Complexity

Let us consider equations involving polynomials $p(y_1, \ldots, y_n)$ and solutions where every variable y_i is 0 or 1. Given such a p, here are several questions we can ask about it.

(a) Is $p(0, \ldots, 0) = 0$?

(b) Does there exist a solution $\boldsymbol{a} \in \{0,1\}^n$ such that $p(\boldsymbol{a}) = 0$?

(c) Are all assignments solutions to the equation $p(\boldsymbol{y}) = 0$?

(d) Are over half of the assignments $\boldsymbol{a}$ solutions?

(e) How many solutions are there?

We can also pose these questions within certain contexts, such as when the following **promise condition** is known to hold in advance:

(f) Either at least 75% of the $\boldsymbol{a}$ are solutions or at most 25% of them are.

In standard presentations of computational complexity, the following is a theorem based on some model-specific definition of the classes. We will instead adopt it as a definition giving a shortcut to formulations that are most useful for framing quantum algorithmic power.

DEFINITION 16.3 A language L or function f belongs to the stated complexity class if there is a classically feasible function g such that for all n and $x \in \{0,1\}^n$, $g(x)$ produces a polynomial $p(y_1, \ldots, y_m)$ such that:

(a) P: $x \in L \iff$ the answer to (a) is yes.

(b) NP: $x \in L \iff$ the answer to (b) is yes.

(c) co-NP: $x \in L \iff$ the answer to (c) is yes.

(d) PP: $x \in L \iff$ the answer to (d) is yes.

(e) #P: $f(x) =$ the number of solutions.

(f) BPP: $x \in L \iff$ the answer to (d) is yes, where (f) holds for all x.

The freedom to choose the **reduction function** g allows some manipulation of equations, such as adding dummy variables or making terms that force certain arguments to certain values in order for a solution to be possible. Doing so shows relations between the questions and hence the classes. For instance, question (a) can be transformed to a case of (c) upon replacing p by $p' = p \cdot (1 - y_1)(1 - y_2) \cdots (1 - y_m)$. Then $p(0, \ldots, 0) = 0$ if and only if all binary assignments $\boldsymbol{a}$ make $p'(\boldsymbol{a}) = 0$. A similar idea transforms (a) into (b), so we conclude:

$$\mathsf{P} \subseteq \mathsf{NP} \cap \mathsf{co\text{-}NP}.$$

That NP $\subseteq$ PP is a bit trickier but uses dummy variables $\boldsymbol{z}$ to make any solution y for $p(\boldsymbol{y}) = 0$ the feather that tips the scales making over half the assignments to $p'(\boldsymbol{y}, \boldsymbol{z})$ be solutions. Because flipping $p(\boldsymbol{y})$ to be $1 - p(\boldsymbol{y})$ flips the answers to questions (a) and (d), the classes P and PP are closed under complementation of their member languages, and it follows also that co-NP $\subseteq$ PP. Given any language $L \in$ NP, these tricks create a reduction function g' such that for all x, $x \in L \iff g'(x) = p'$ belongs to the language L_d of polynomials p' for which over half the assignments $\boldsymbol{a}$ make $p'(\boldsymbol{a}) = 0$. This is summarized by saying that L_d is NP-**hard**. We have defined things so that the language L_b of p for which there exists a solution is immediately NP-hard, and because L_b also belongs to NP, it is **NP-complete**.[1]

That BPP $\subseteq$ PP is immediate by relaxing the promise condition (f), and BPP is likewise closed under complement. Because (f) is maintained in the trivial reduction to (d) from (a), we have P $\subseteq$ BPP, but the methods used in going from (b) or (c) to (d) do not preserve it, and neither NP nor co-NP is known to be contained in BPP.

Questions about whether there are at least k solutions can be tweaked into the form (d) via dummy variables. Binary search on whether there are at least k solutions then enables feasibly counting their number, so question (e) (which clearly subsumes the other questions) is roughly equivalent to (d). Technically, one cannot compare PP to #P directly, because the latter is a function class, and getting multiple values of the form (e) might be more powerful than a single question (d). However, asking for the *difference* between the numbers of solutions and nonsolutions stays within the power of (d), a fact we use to conclude BQP $\subseteq$ PP. For now we note:

THEOREM 16.4 BPP $\subseteq$ BQP. □

The polynomial equations that arise in this development have other uses for analyzing quantum circuits, and we explore them next.

16.3 A Circuit Labeling Algorithm

We give an algorithm for labeling a quantum circuit algebraically. The end result is a polynomial equation for which the difference of two numbers of solutions yields the circuit's acceptance probability. We first give a form in which the polynomials have values +1 and −1, which correspond to the signs in the sum-over-paths explication we gave for quantum measurement and effects in chapter 7. This uses the value 0 to eliminate impossible paths. Then we show in section 16.5 how to make do without the value −1 and reduce the degrees of the equations drastically.

1 With respect to a machine-model definition of NP this is a theorem, indeed an offshoot of the fundamental **Cook-Levin theorem** as explored in this chapter's exercises. We remark also that in all these cases, we can obtain a feasible function g'' that first builds a polynomial $p(x_1,\dots,x_n,y_1,\dots,y_m)$ that depends only on n, and then obtains $p_x = p(y_1,\dots,y_m)$ by substituting the actual value of each bit x_i of the given x for the corresponding formal variable x_i.

The algorithm goes in stages for each new gate working left to right, that is, from what we regard as inputs to outputs. It assumes the wires into a gate have already been labeled, and it labels the gate's outgoing wires, with unaffected qubit wires keeping their labels. By theorem 16.2 we can restrict attention to Hadamard and Toffoli gates, although we include ***CNOT*** because it is useful to show, and the exercises treat how to extend this for other quantum gates. We let u_i, u_j, u_k stand for the current labels on the qubits i, j, and/or k involved in a gate.

1. Label the inputs with variables $x_1, \ldots, x_n$. If there are ancilla qubits, then continue labeling them $x_{n+1}, \ldots, x_m$, although if they will always be initialized to 0, then one can label them 0 straightaway.
2. Label the outputs with variables $z_1, \ldots, z_n$, again using more if there are more qubits. From here on we will not need to address ancilla qubits as special and will just say "n" for the end index.
3. Let h be the number of Hadamard gates in the circuit, and allocate variables $y_1, \ldots, y_h$.
4. Initialize a polynomial P, called the *global phase polynomial*, to the constant 1.
5. For the next Hadamard gate $\boldsymbol{H}_j$ on some qubit line i, allocate the fresh variable y_j, multiply P by the factor $(1 - 2u_i y_j)$, and make y_j the new label on line i.
6. For a ***CNOT*** gate, leave the control label u_i unchanged, but change u_j to $u_j + u_i - 2u_i u_j$. There is no change to P.
7. For a Toffoli gate with controls on lines i, j, leave u_i and u_j alone, but change the target u_k to $u_k + u_i u_j - 2u_i u_j u_k$. There is no change to P.
8. When done with all the gates, for each i, create the *measurement constraint* $e(u_i, z_i)$, where u_i is the last label on line i and

   $$e(u, z) = 2uz + 1 - u - z.$$

   Note that $e(0,0) = e(1,1) = 1$, whereas $e(1,0) = e(0,1) = 0$, so these enforce equality of the final labels and the outputs on Boolean outcomes. The final polynomial $R = R_C$ is the product of P and all the measurement constraints.

We may instead consider the measurement constraints to be defined at the beginning as $e(x_i, z_i)$ for all i, and whenever a label u_i is changed to v_i, the constraint $e(u_i, z_i)$ is changed to $e(v_i, z_i)$. Here are two examples of the labeling algorithm. The first puts something in place of the "?" label in the circuit example from section 7.6, which showed entanglement:

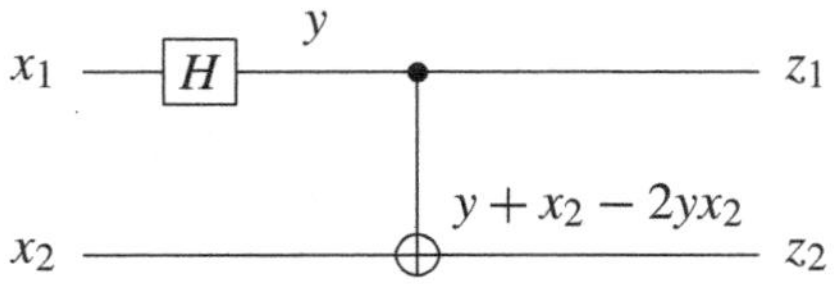

x_1 — H — y — z_1

x_2 — $y + x_2 - 2yx_2$ — z_2

(16.2)

$$
\begin{aligned}
P &= 1 - 2x_1y \\
R &= P \cdot e(y, z_1)e(y + x_2 - 2yx_2, z_2) \\
&= (1 - 2x_1y) \cdot (2yz_1 + 1 - y - z_1) \\
&\quad \cdot (2(y + x_2 - 2yx_2)z_2 + 1 - y + x_2 - 2yx_2 - z_2).
\end{aligned}
$$

When we substitute the input 00, that is, $a_1 = 0$ for x_1 and $a_2 = 0$ for x_2, this simplifies to $P = 1$ and $R = e(y, z_1)e(y, z_2)$. We can infer from this that the only allowed output states have $z_1 = z_2$, which shows the entanglement.

The following larger example is adapted from two sources that treated the output constraints as separate equations:

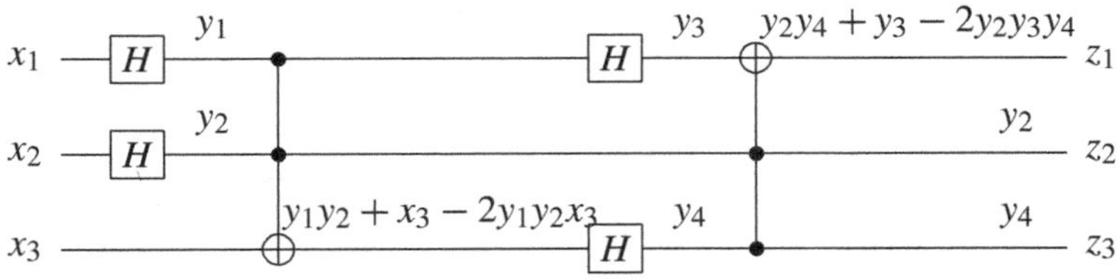

x_1 — H — y_1 — H — y_3 — $y_2y_4 + y_3 - 2y_2y_3y_4$ — z_1

x_2 — H — y_2 — y_2 — z_2

x_3 — $y_1y_2 + x_3 - 2y_1y_2x_3$ — H — y_4 — y_4 — z_3

$$
\begin{aligned}
P &= (1 - 2x_1y_1)(1 - 2x_2y_2)(1 - 2y_1y_3)(1 - 2y_4(y_1y_2 + x_3 - 2y_1y_2x_3)) \\
R &= P \cdot e(y_2y_4 + y_3 - 2y_2y_3y_4, z_1)e(y_2, z_2)e(y_4, z_3).
\end{aligned}
$$

16.4 Sum-Over-Paths and Polynomial Roots

Now let $\boldsymbol{U_1U_2 \cdots U_s}$ be the matrix representation of a quantum circuit, that is, as a product of $N \times N$ matrices, and let $a, b \in [N]$. Consider a path

$$\boldsymbol{U_1}[a, c_1]\boldsymbol{U_2}[c_1, c_2] \cdots \boldsymbol{U_{s-1}}[c_{s-2}, c_{s-1}]\boldsymbol{U_s}[c_{s-1}, b].$$

Call the path *positive* if it multiplies out to $+1$, *negative* if it multiplies out to -1, and *zero* if one of the entries is 0, which is the only other possibility for circuits of Hadamard, ***CNOT***, and Toffoli gates. Let $p^+(a,b)$ be the number of positive paths from a to b and $p^-(a,b)$ the number of negative paths. We will again use the "Phil the mouse" visualization of these paths—recall that the "maze gadgets" for these gates were also depicted in chapter 7. In this section we will conserve these quantities individually, not just their difference $p^+(a,b) - p^-(a,b)$, which gives the amplitude of "surviving Phils."

Finally, referencing the polynomial R, define $N_R[+1 \mid a; b]$ to be the number of assignments y to $y_1, \ldots, y_h$ that make $R(a; y; b) = 1$. Define $N_R[-1 \mid a; b]$ similarly for $R(a; y; b) = -1$. Here $R(a; y; b)$ means we are substituting a for $x_1, \ldots, x_n$ and b for $z_1, \ldots, z_n$.[2]

The following technical lemma connects the sum-over-paths formulation with the numbers of solutions to the equations $R = 1$ and $R = -1$:

LEMMA 16.5 For all circuits C on n qubits as above, and $a, b \in \{0,1\}^n$,

$$\begin{aligned} p^+(a,b) &= N_R[+1 \mid a; b], \\ p^-(a,b) &= N_R[-1 \mid a; b]. \end{aligned} \tag{16.3}$$

Proof. We work inductively as C is built gate-by-gate from an initially empty circuit C_0. C_0 has one positive path from a to a for each a, none from a to b when $b \neq a$, and no negative paths anywhere. The initial polynomial R_0 is

$$\prod_{i=1}^{n} e(x_i, z_i)$$

because there are no y variables. Whenever $a = b$, $R_0(a,b) = 1$, whereas $a \neq b$ means $a_i \neq b_i$ for some i, whereupon $e(a_i, b_i)$ zeros out the product. We technically satisfy $N_R[+1 \mid a; a] = 1$ for each a because there is exactly one $y \in \{0,1\}^0$, namely, the empty string, whereas $N_R[-1 \mid a; a] = 0$ because nothing gives a value of -1. Thus, the lemma's properties hold for C_0.

For the induction, let C, P, R satisfy the equalities in the lemma, and suppose we obtain a new circuit C' first by adding one Hadamard gate on qubit line i. Let u denote the label before the gate on that line. Let us fix an input a and output b except for the value b_i on that line. That is, we consider the two outputs $b[b_i = 0]$ and $b[b_i = 1]$. Let p_0^+, p_0^- stand for the numbers of paths of C from a to $b[b_i = 0]$ that multiply out to $+1$ and -1, respectively. Write p_1^+, p_1^-

2 The exercises explore ideas such as if one intends only to measure the first qubit, then one may delete the measurement constraint factors other than $e(u_1, z_1)$, or alternatively one may leave outputs other than z_1 unsubstituted and e.g. define $N_R[-1 \mid a; b_1]$ to be the number of assignments to $y_1, \ldots, y_h$ and $z_2, \ldots, z_n$ that yield -1 when b_1 is substituted for z_1 only.

similarly for the case $b_i = 1$, and let $q_0^+, q_0^-, q_1^+, q_1^-$ denote the corresponding quantities in the new circuit C'.

Now we can visualize the maze gadgets for Hadamard gates from chapter 7, which here have a -1 path from $b[b_i{=}1]$ for C to the same terminal for C' and three positive paths involving the $b[b_i{=}0]$ terminals. The maze corridors simply carry the values of the Hadamard matrix applied to line i, so we have:

$$\begin{aligned} q_0^+ &= p_0^+ + p_1^+ \qquad & q_1^+ &= p_0^+ + p_1^- \\ q_0^- &= p_0^- + p_1^- \qquad & q_1^- &= p_0^- + p_1^+. \end{aligned} \tag{16.4}$$

In terms of the new polynomials P' and R' for C', we have $P' = P \cdot (1 - 2uy_h)$, and R' replaces $e(u, z_i)$ with $e(y_h, z_i)$. With a fixed, let S_0 denote the set of assignments to $y_1, \ldots, y_{h-1}$ that make $u = 0$ and make P have value $+1$. Let T_0 similarly stand for $u = 0$ and P having the value -1, and S_1, T_1 likewise for $u = 1$. Now let S_0' instead denote the assignments to $y_1, \ldots, y_h$ with $y_h = 0$ that give P' the value $+1$ and T_0' those with $y_h = 0$ that give -1; note that these arise only for $b_i = 0$. Finally, let S_1', T_1' denote the corresponding sets with $y_h = 1$. An assignment in S_0' is free to make u have either value because $y_h = 0$ makes P' have the same values as P, and applying this to the case $b_i = 0$ takes care of the $e(y_h, z_i)$ term. The similar observation for T_0' gives us likewise a 1-to-1 correspondence, and with a slight abuse of notation because $y_h = 0$ is fixed, we write them as disjoint unions:

$$\begin{aligned} S_0' &= S_0 \uplus S_1 \\ T_0' &= T_0 \uplus T_1. \end{aligned}$$

Now for the case $b_i = 1$, we need $y_h = 1$, and this flips the sign of assignments that also make $u = 1$. We therefore obtain:

$$\begin{aligned} S_1' &= S_0 \uplus T_1 \\ T_1' &= T_0 \uplus S_1. \end{aligned}$$

By the properties for C, we have $p_0^+ = ||S_0||$, $p_0^- = ||T_0||$, $p_1^+ = ||S_1||$, and $p_1^- = ||T_1||$. Substituting these into the right-hand sides of (16.4) yields the goal identities for C' that correspond to (16.3) for C.

In the case of a ***CNOT*** (respectively, Toffoli) gate with target on line i and control(s) on line j (and k), fix any a, b, and let b' have $b_i' = b_i \oplus b_j$ (respectively, $b_i' = b_i \oplus b_j b_k$). It is incidental but helpful to note in both cases that the map from b to b' is its own inverse. The last "maze stage" for C' shunts paths of C ending at b to b', with no branching or sign change, and vice versa. Hence, with similar notation to before, we have:

$$q_{b'}^{+} = p_b^{+};$$
$$q_{b'}^{-} = p_b^{-}.$$

By the induction hypothesis, p_b^{+} is equal to the number of assignments to $y_1, \dots, y_h$ that make P have value 1, that is, to $N_R[+1 \mid a; b]$. Similarly, $p_b^{-} = N_R[-1 \mid a; b]$. We have $P' = P$ and no new variable, so the only difference is that $e(u, z_i)$ in R is replaced by $e(u \oplus v)$ (respectively, $e(u \oplus vw)$) in R', where u is the previous label on line i, and v, w are the unchanged labels on the control(s). The changes make

$$N_R[+1 \mid a, b] = N_{R'}[+1 \mid a; b'];$$
$$N_R[-1 \mid a, b] = N_{R'}[-1 \mid a; b'].$$

Because the left-hand sides are equal to p_b^{+} and p_b^{-}, respectively, the right-hand sides are equal to $q_{b'}^{+}$ and to $q_{b'}^{-}$, as needed to be proved. □

As is already evident from the circuit examples in section 16.3, the multiplication makes the degree of P ramp up as gates are added. The number of terms ramps up even more if the product is multiplied out. We use one more wrinkle to make the terms add rather than multiply.

16.5 The Additive Polynomial Simulation

This is a short section but gives in some sense the tightest classical rendition of nature's quantum computing power. It maps into the additive structure of integers modulo $k = 2$ rather than the multiplicative structure of $+1, -1$. The new wrinkle is that if an equality constraint $e(u_i, z_i)$ is going to be violated, then let us allocate a fresh variable w_i and add the term

$$w_i(1 - e(u_i, z_i)) = w_i(u_i + z_i - 2u_i z_i),$$

which further becomes simply $w_i(u_i + z_i)$ under addition modulo 2. Consider now assignments Z to all the $x; y; z$ variables that make $u_i \neq z_i$. The multiplier of w_i becomes $+1$, so the final R will have the form $R' + w_i$, with w_i appearing nowhere else in R. For every assignment that violates the constraint, one value of w_i will give 1 and the other will give 0, so they will cancel out with regard to the difference

$$N_R[0 \mid a; b] - N_R[1 \mid a; b].$$

Now in place of P we initialize an additive phase polynomial Q to 0, and the only other changes we need to make to the labeling algorithm are:

1. For a Hadamard gate on line i, with u_i and y_h as before, **add** to Q the term $u_i y_h$.
2. The label change on a ***CNOT*** gate simplifies to $u'_i = u_i + u_j$, with again no change to Q.
3. For a Toffoli gate with controls on lines i,j and target on line k, the new target label is $u'_k = u_k + u_i u_j$.
4. At the end, $R = Q + \sum_i w_i(1 - e(u_i, z_i))$.

For example, in the simple Hadamard + ***CNOT*** circuit diagram (16.2) in section 16.3, the label $y + a_2 - 2ya_2$ becomes simply $y + a_2$. The polynomials of the larger circuit diagram in section 16.3 become:

$$\begin{aligned} Q &= x_1 y_1 + x_2 y_2 + y_1 y_3 + (y_1 y_2 + x_3) y_4; \\ R &= Q + w_1(y_3 + y_2 y_4 + z_1) + w_2(y_2 + z_2) + w_3(y_4 + z_3). \end{aligned}$$

LEMMA 16.6 For all quantum circuits C on n qubits as above and $a, b \in \{0,1\}^n$,

$$p^+(a,b) - p^-(a,b) = N_R[0 \mid a; b] - N_R[1 \mid a; b].$$

Proof. The verification details are a direct carryover from the proof of lemma 16.5, except that the presence of the w_i variables implies equivalence only for the difference $p^+ - p^-$, not for the individual $+$ and $-$ path counts as before. □

16.6 Bounding BQP

Our upper bound on BQP follows quickly from the characterization of quantum circuits by Hadamard and Toffoli gates in theorem 16.2. We restate the lower bound from theorem 16.4 as well.

THEOREM 16.7 BPP $\subseteq$ BQP $\subseteq$ PP.

Proof. Given feasible quantum circuits C for a BQP algorithm with h-many nondeterministic (i.e., Hadamard) gates, we may assume that on any input $x \in \{0,1\}^n$, C starts with $x0^{m-n}$ and is measured so that the result $b = 10^{m-1}$

gives acceptance. This uses results in chapter 6, especially section 6.3. By lemma 16.5, we quickly obtain a polynomial $R(x; y; z)$ such that upon substituting $a = x0^{m-n}$ for x and b for z, the acceptance probability is the square of the difference between $N_R[{+1} \mid a; b]$ and $N_R[-1 \mid a; b]$, divided by 2^h. (Or by lemma 16.6, we obtain $R(w; x; y; z)$ such that the probability is $(N_R[0 \mid a; b] - N_R[1 \mid a; b])^2$ divided by 2^{h+2m} where m is the number of w_i variables employed.) Computing this difference exactly can be done in PP, and this suffices to distinguish whether the circuit accepts or rejects a given input. □

Getting the exact difference is actually overkill because a coarse approximation d need only separate the case $d^2/2^h > 2/3$ from $d^2/2^h < 1/3$. Note, however, that the approximations must be within some $(1 + \epsilon)$ factor of d, not just within such a factor of $N_R[{+1} \mid a; b]$ and $N_R[-1 \mid a; b]$ individually. Because $N_R[{+1} \mid a; b] - N_R[-1 \mid a; b]$ is divided by $\sqrt{2^h}$, not by 2^h, the two will always come closer to canceling than a fixed-factor approximation to either can distinguish.

No containment relation is known between BQP and NP. In particular, although counting solutions of polynomials is NP-hard, and although this theorem implies that counting solutions is enough to determine whether a quantum circuit will accept, this does not mean that quantum circuits can solve NP-hard problems. Claims to this effect have been notorious. More notable, however, have been efforts to harness quantum processes to find heuristic solutions in many cases that are perhaps indelibly harder to obtain classically. Hence, there is interest in the possible heuristic solvability of some of the resulting problems in polynomial algebra.

16.7 Problems

16.1. Suppose f and g are feasible functions and L is a language such that for all x, $g(x)$ is a polynomial $p(y_1, \dots, y_m)$ such that $x \in L \iff N_p[1] - N_p[0] \geq f(x)$. Prove that L is in PP.

16.2. Consider a NAND gate g with inputs u and v, and any of its output wires w. Write a Boolean formula ϕ in literals $\pm u, \pm v, \pm w$ that is satisfied by exactly those assignments for which w is the correct output of NAND(u, v).

16.3. Suppose C is a classical circuit of NAND gates with inputs $x_1, \dots, x_n$ *and* $y_1, \dots, y_m$ and output wire w_o. Show how to construct a Boolean formula ϕ_C

that is satisfiable exactly when there exist $x \in \{0,1\}^n$ and $y \in \{0,1\}^m$ such that $C(xy)$ outputs 1. How many extra variables do you need in terms of wires in C?

16.4. Conclude that for any uniform family of circuits C_n, the language L defined for all x by

$$x \in L \iff (\exists y)\ C_n(xy) = 1$$

feasibly reduces to the language SAT of Boolean formulas ϕ that are satisfiable. Note that you can take ϕ_C from problem 16.3 and substitute for the particular bit-values of x. This yields the **Cook-Levin theorem** that SAT is NP-complete.

16.5. Show further that ϕ_C and ϕ can be written as conjuncts of *clauses*, where each clause is a single literal (such as w_0 or $\pm x_i$), a disjunction of two positive literals, or a disjunction of three negative literals. This is a form of the problem called 3SAT, which is likewise NP-complete.

16.6. Now convert the ϕ in problem 16.2 into a polynomial $p(u,v,w)$ such that $p(a_1,a_2,a_3) = 0$ for exactly those assignments $a_1a_2a_3$ that satisfy ϕ. What is its degree? Note that a NAND gate is simulated on binary arguments u, v by the polynomial $1 - uv$, but what's more important than translating the gate is translating the specification that the gate is working correctly.

16.7. Build on the last problem to show a reduction from SAT (or alternately from 3SAT) to the solution-existence problem for polynomials, which is question (b) in section 16.2. Formally, this makes these two problems meet the definition given there of belonging to NP, so that they are NP-*complete*.

16.8. Conclude from all this that if you define NP initially via polynomial-time nondeterministic machines or circuits, then the solution-existence problem is thereby shown to be NP-complete.

16.9. Suppose we add a ***NOT*** gate to a quantum circuit. Work out the change to labels and the polynomial P needed to make the algorithm and verification of lemma 16.5 hold for this gate as well.

16.10. Same as the last exercise, except now we add a ***CNOT*** gate.

16.11. Suppose we have a kind of gate ***W*** for which we have an associated change of labels on the qubit lines it affects and update factors for the polynomial P (or rather Q). Show how to derive the labels and factors for the ***CW*** (i.e., Controlled-***W***) gate based on this information alone.

16.12. Add a case to the labeling algorithm and to the proof of lemma 16.5 for a ***NOT*** gate.

16.13. Complete the case of the proof of lemma 16.5 for a ***CNOT*** gate.

16.14. Give the polynomials P_C and Q_C for the trivial one-qubit circuit with two Hadamard gates:

x — H — y — H — z

Show, using R_C as well, that this circuit is equivalent to the identity for all measurements.

16.15. Now consider again the circuit from section 6.3:

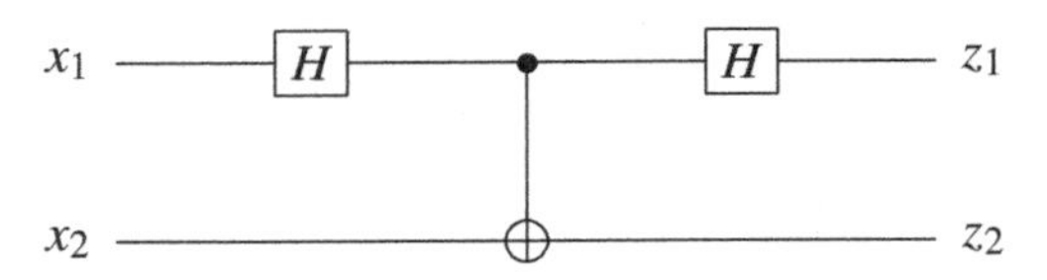

Show that the conclusion of problem 16.14 does *not* hold for the first qubit by showing nonzero amplitude difference for case(s) with $z_1 \neq x_1$. This shows that the inclusion of equality constraints in R_C compared with P_C or Q_C matters in the equations. Use R_C to compute the amplitudes of all outcomes on input $x = 00$.

16.16. Suppose we introduce a so-called ***CZ*** gate, which incidentally is the Grover oracle marking the string 11:

$$\boldsymbol{CZ} = \begin{bmatrix} 1 & 0 & 0 & 0 \\ 0 & 1 & 0 & 0 \\ 0 & 0 & 1 & 0 \\ 0 & 0 & 0 & -1 \end{bmatrix}.$$

What happens for this gate in the labeling algorithm and the proof of lemma 16.5?

16.17. Modify the definition of R_C and/or the proof of theorem 16.7 for the regime where we measure only the value of the first qubit.

16.18. Modify the additive "Q version" of the labeling algorithm to allow the phase polynomial to have values $0, 1, 2, 3$ in the integers with addition modulo 4, which gives a logarithmic representation of the multiplicative structure of the values $1, i, -1, -i$. Use this to incorporate into the labeling algorithm the so-called $\boldsymbol{S}$-gate defined by

$$\boldsymbol{S} = \begin{bmatrix} 1 & 0 \\ 0 & i \end{bmatrix}.$$

The analogous version of lemma 16.6 becomes more complicated but retains the essence that the acceptance probability is a simple function of the values $N_R[k \mid a; b]$ for $k = 0, 1, 2, 3$.

16.19. Observe that when there are no Toffoli gates, that is, when only the gates mentioned in the above problems plus Hadamard gates are used, the polynomials Q obtained have degree no higher than two. Use the known theorem that solution-counting is in deterministic polynomial time for such polynomials over the integers modulo 4 to conclude the **Gottesman-Knill theorem**: every so-called *stabilizer circuit* of $\boldsymbol{H}, \boldsymbol{S}, \boldsymbol{CZ}$ gates only (plus $\boldsymbol{X}$, $\boldsymbol{CNOT}$, and some others) of size s can be simulated in classical polynomial time $s^{O(1)}$.

16.20. Let us extend the additive polynomial simulation to circuits with $\boldsymbol{T}$-gates to work over $\mathbb{Z}_8$ as follows: For a new Hadamard gate on a qubit line with label u, allocate the fresh variable y_j as the new label and add $4uy_j$ to the phase polynomial Q. (The general rule modulo 2^r is to add $2^{r-1}uy_j$.) For a $\boldsymbol{T}$-gate, leave the label alone but add u^2 to Q. State and prove the appropriate version of lemma 16.6 for this representation. (Thus, the "Crescent Phils" and "Gibbous Phils" mentioned in section 7.5 are formally taken care of here.)

16.21. With reference to problem 16.20, analyze the one-qubit commutator circuit formed by $\boldsymbol{HTHT}^{-1}$. What are the probabilities of measuring 0 and 1 on input $\boldsymbol{e}_0$ and $\boldsymbol{e}_1$, respectively?

16.8 Summary and Notes

The accepted definition of BQP comes from the appropriately named 1997 article "Quantum Complexity Theory" by Bernstein and Vazirani (1997). Its polynomial-space upper bound was improved to PP by Adleman et al. (1997). The sharper observation that it belongs to a definitionally smaller class (called AWPP) characterized by taking the difference of two #P functions was set down by Fortnow and Rogers (1999). Our proof of this follows Dawson et al. (2004), while the example preceding lemma 16.5 comes also from Gerdt and Severyanov (2006). A different, novel, and informative proof of BQP $\subseteq$ PP via the concept of "post-selection" comes from Aaronson (2005), and this also simplified the proof of known closure properties of PP. For the Solovay-Kitaev theorem we refer to Dawson and Nielsen (2006). The Gottesman-Knill theorem was first set down by Gottesman (1998), and the first of several improvements to the algorithm came from Aaronson and Gottesman (2004). The theorem about counting solutions of quadratic polynomials is joint work of this text's first author (Cai et al., 2010), while the relevance to Gottesman-Knill was noticed by the second author. Much further work of high interest is due to Aaronson (2010) and Aaronson and Arkhipov (2010); see also the recent popular book by Aaronson (2013).

17 Beyond

Our goal has been to get you started understanding some of the basic concepts from quantum algorithms. We hope that you now have a working understanding of them.

One of the takeaways is that the quantum algorithms are not that complicated, nor are they based on mysterious ideas. They are not even necessarily complex, neither in the sense of needing complex numbers nor in having many loops and high running times. We hope that you now see that quantum algorithms in general have a simple structure and that their analysis is quite standard. The proofs that show they work use basic tools from linear algebra and number theory—nothing strange or exotic is required.

17.1 Reviewing the Algorithms

For each quantum algorithm, we have shown a respect in which it surpasses a classical algorithm for the same task. This started by counting the number of evaluations of a function f given as a black-box parameter of the task. With Deutsch's algorithm, we saved one evaluation of f in the worst case by using the ability in the quantum world to apply f to a linear combination of the basis vectors denoting argument strings. Then with the Deutsch-Jozsa algorithm, we saved more such queries, and with Simon's algorithm, we saved exponentially many even in the expected case. Shor's algorithm surpassed in raw time any known classical algorithm, while Grover search and quantum walk search algorithms confer a definite polynomial savings.

The first question to consider, going beyond these algorithms, is what is the common origin of these savings? What is the quantum core of them? We offer the following candidate for "the" common feature:

> Quantum computation is a great **amplifier** of slight regularities and deviations.

In a Grover search, the initial deviation is as small as possible—it can be just one solution that flips the sign of one component in an exponential-sized start vector. This gives the minimum improvement: $O(2^{n/2})$ iterations in place of $\Theta(2^n)$ expected time. In quantum walks, the linear time for spreading (compared with quadratic time for classical walks on undirected graphs) preserves the amplification in the simple Grover setting and focuses the mechanism of amplification for greater savings when the solution set size can be inflated while keeping small time for updates between evaluations of the search function. The greater savings in Shor's algorithm can then be viewed as arising because *periodicity* is a greater regularity than the presence of a solution. It was still far from obvious that a polynomial amount of work would suffice to amplify it, but Simon's algorithm provided a hint. In Simon's algorithm, the reason a linear amount of work suffices with high probability is that $2n$ random choices usually suffice to find a basis of n independent binary vectors in n-dimensional Hilbert space, even though the set of all binary vectors has exponential size. This goes also for related *hidden subgroup* problems, which we have passed over, except for a mention in problems 10.6–10.8, and which constitute one possible line of further study.

What will it take to find a great *new* kind of quantum algorithm? It may need first thinking up a new kind of amplification.

17.2 Some Further Topics

To sum up the various paths from this point on, all of the following can be amplified for further study without canceling each other out. There is much more to learn from what is already known and much more to learn that is being discovered right now.

• **More about algorithms.** All of the algorithms we have covered, especially from Simon's algorithm onward, have further developments. For example, some new results by Belovs et al. (2013) and Jeffery et al. (2013) have come from the idea of *nesting* quantum walks in a *quantum* manner, not just classically as with the recursion we sketched for the chess/NAND-tree example in chapter 15. Movement toward regarding quantum routines more as enabling *sampling* from classically difficult distributions is also current.

• **More about computational complexity.** We have only scratched the surface of complexity classes neighboring BQP whose relationships to each other, let alone to quantum classes, form one of the great mysteries of our age. Besides whether BQP $\neq$ BPP and whether factoring is in the difference, there are the questions of whether BQP contains the graph isomorphism problem and whether it is contained in any level of the *polynomial hierarchy* above NP.

• **More about quantum physics.** There are a plethora of possibilities. Before you go any further, you will need to internalize Dirac's bra-ket-etc. and other physics notation. Then it will help to study evolution according to Schrödinger's equation. This will open the way to other physical models, such as adiabatic quantum computation. A further physical topic is topological quantum computation. With all of these models, there is the question of building physical quantum computers. We hosted a year-long debate on this issue between Gil Kalai and Aram Harrow on the *Gödel's Lost Letter* blog in 2012.

• **More about quantum implementations and circuits.** There is a huge literature on both these related topics that you can read more about. This will explain and show in greater detail why all the unitary transforms we called feasible indeed are. Other longer, more standard textbooks than ours are places to start. There are many new papers and expositions on the engineering challenges every month. The most important topic not included in this text is the **quantum fault-tolerance theorem** and its beautiful use of error-correcting codes, as may be found in all the larger texts we have referenced.

• **More about quantum information theory, communication, and cryptography.** They go together, are included in all of the longer texts, and provide applications here and now. We covered some of the basics in section 8.3 and mentioned Holevo's theorem that transmitting an n-qubit quantum state in isolation can confer at most n bits of classical information. The idea of quantum cryptography is to build new types of security systems that rely not on the hardness of some computational problem—like classical cryptography does—but on the special behavior of quantum systems. Actual usable physical devices have been built and put into service to secure the most sensitive financial transactions.

• **More about related mathematical and scientific areas.** Shor's algorithm certainly shows how several areas of *number theory* are needed to channel the quantum outputs. Quantum computing may be a coal-mine canary for the prospects of various "theories of everything" in science. Biological processes from DNA upward are already profitably regarded as computational, and one can expect quantum imprints. The title and blurb of the article by Ball (2011) say it all: "Physics of life: The dawn of quantum biology. The key to practical quantum computing and high-efficiency solar cells may lie in the messy green world outside the physics lab."

Specific to linear algebra, we mention two nifty theorems from the beautiful theory of spectral decompositions:

1. Every self-adjoint matrix $\mathbf{A}$ on $\mathbb{C}^n$ has an orthonormal basis of eigenvectors $\mathbf{x}_i$ with real eigenvalues λ_i, such that $e^{i\mathbf{A}}$ is well defined (using the same Dirac notation for projectors via outer products that was discussed in section 6.4) by:
$$e^{i\mathbf{A}} = e^{i\lambda_1}|\mathbf{x}_1\rangle\langle\mathbf{x}_1| + \cdots + e^{i\lambda_n}|\mathbf{x}_n\rangle\langle\mathbf{x}_n|.$$
2. Every unitary matrix $\mathbf{U}$ arises as $\mathbf{U} = e^{i\mathbf{A}}$ for some self-adjoint $\mathbf{A}$ that is also unitary, so that each λ_i is a unit complex number identified with an angle.

This opens the way to understanding things like Schrödinger's equation and the role of Planck's constant (called h or $\hbar = \frac{h}{2\pi}$) in topics like the **Heisenberg Uncertainty Principle**. These are advanced physics topics, but a nonphysicist can draw on linear algebra and ideas of computations to approach them. For instance, the general solution to Schrödinger's equation in the case of a time-independent Hamiltonian operator $\mathbf{H}$ is

$$\mathbf{U}(t) = e^{-i\mathbf{H}t/\hbar}.$$

Here $\mathbf{H}$ does not stand for Hadamard, but, like Hadamard transforms, the Hamiltonian operators are *self-adjoint*, so we can apply the above to understand the whole right-hand side as a unitary operator for any t, which is what the left-hand side says it is. Now you can see ways in which this is like a computation. Then we can even ask questions like how far all this is analogous to the way our additive simulation in section 16.5 works with the logarithms of the quantities in the multiplicative simulation of section 16.4.

There are many more topics beyond this book than we can list. One general phenomenon is that many important concepts in classical computation have been reissued in "quantum" versions. Besides BQP being the quantum version of BPP, there is a quantum analogue QNP of NP and so on for many other complexity classes. There are quantum finite automata, quantum cellular automata, and even quantum formal grammars. Quantum communication complexity and information assurance has evolved into a big field, with both quantum versions of classical protocols and distinctively quantum situations. We end, however, by looking inward at what is distinctively "quantum" about the algorithms we have studied, quantum particularly meaning beyond the known reach of classical computations.

17.3 The "Quantum" in the Algorithms

We began with the question: can we build a physical quantum computer to confer the promised advantages of quantum algorithms? We have not covered the engineering challenges in this text or the quantum fault-tolerance theorem. That theorem sets concrete rather than asymptotic physical thresholds for making quantum error-correction work, which is a prerequisite for enabling current designs of quantum computers to be engineered. Before we get there, however, we can pose a simpler question based on what we have learned:

Where do the advantages of quantum algorithms come from?

Quantum mechanics gives a new *lever* for algorithms. Archimedes famously said, "Give me a place to stand, and I shall move the Earth"—meaning a place far enough away for a long enough lever. His Greek words did not say, however, where he would find a *base* for the lever. With quantum states, we certainly have exponentially long vectors, but what is the base—where does the power reside? We consider some possible questions and answers:

• Is the power already in the exponentially long notation? *Not in terms of information*—by Holevo's theorem mentioned above, transmitting an n-qubit quantum state can confer at most n bits of classical information. This is so even if the state is a feasible relational superposition $\boldsymbol{s_r} = \sum_{x,y:\ R(x,y)} \boldsymbol{e}_x \otimes \boldsymbol{e}_y$ (suitably normalized). This appears to have information about values related to every x, but the state allows extracting only one such entire value via measurement. More precisely, it yields at most $|x| + |y|$ bits of information. This limit also matters if we try to encode a general n-vertex graph with $\Theta(n^2)$ edges using only n qubits. We wonder which relations R other than the graphs of feasible functions may make $\boldsymbol{s_r}$ feasible to prepare.

• Is it in the ability to generate entanglement? *Not alone.* Although Hadamard and ***CNOT*** gates can generate lots of entanglement and together with the ***S***-gate can create all the error-correcting codes needed for the fault-tolerance theorem, nevertheless by the Gottesman-Knill theorem outlined in the exercises of chapter 16, circuits of these gates all have feasible classical simulations. Measures of entanglement, especially for n-partite not just binary systems, have no agreed-on standard definition, while some working definitions have high complexity.

• Does it have to do with interference? *Definitely yes, but...* The "but" is that exponential savings from interference presupposes that we have harnessed exponential power somehow in generating the cancellations, which throws us back into points and discussions above. However, simple, nonexponential interference yields many working applications in quantum communication and can be experienced by anyone with two sheets of polarizing material.

• Is it in superpositions such as $\boldsymbol{j}_n = \frac{1}{\sqrt{N}} \sum_{x \in \{0,1\}^n} \boldsymbol{e}_x$, which despite its exponential size as a formula, is feasibly obtained by applying the Hadamard transform? *At least in part, but...* The "but" this time is that working on $\boldsymbol{j}_n$ is not the same as having true "exponential parallelism." If it were, then we would expect quantum algorithms to be able to solve NP-complete problems by making unstructured (Grover) search run in polynomial time. Not only is there no real evidence for $\mathsf{NP} \subseteq \mathsf{BQP}$, but we have noted that $\Omega(2^{n/2})$ is a lower bound on the number of queries needed by Grover's algorithm specifically and hence on its running time.

• Is it specific to the quantum Fourier transform? *Oddly there is evidence for "no."* First, we have noted, although not proved, that Shor's algorithm requires only moderately coarse approximations to the QFT, and that these are readily obtained via circuits of other fundamental gates, even Hadamard and Toffoli gates alone. These results do presuppose the ability to carry out the n-qubit *Hadamard* transform. Second, because Shor's algorithm does a complete measurement immediately after applying the QFT, the same distribution can be obtained by iteratively measuring each qubit after applying just $\boldsymbol{H}_2$ and $\boldsymbol{R}_\theta$ to it, where the rotation angle θ depends on the results of previous measurements, for which see Preskill (2004). Thus, for factoring, the QFT can "deconstruct" into simple single-qubit operations and measurements. Third, there is also classical super-polynomial power in Simon's algorithm, although (a) the problem it solves does not have as strong evidence as Shor's of being beyond classical reach, and (b) denying the classical solvers for Simon's problem the ability to use field extensions and linear combinations may give an unfair comparison. The fourth line of evidence against the QFT alone being the fulcrum comes next.

• Is it in the functional superpositions? Referencing a famous *Peanuts* cartoon, we are tempted to say, "That's it!" In the cartoon, Charlie Brown is in session at Lucy's psychiatry stand, and she enumerates names of phobias to see which could be troubling him. When she comes to "pantophobia" (the fear of everything), Charlie's exclamation knocks both off their chairs.

There are several senses in which the full QFT matrix $\boldsymbol{F}_N$ has feasible classical simulations, such as when it is given a product state—any product state, not necessarily one of the standard basis states—as input. These simulations are not known to apply to functional superposition states; if they did, then factoring would be in BPP. See Aharonov et al. (2006), Jozsa (2006), Markov and Shi (2008), van den Nest (2013). The last of these references asserts that the power must reside somewhere in the interface between the quantum and classical components of Shor's algorithm, and the superpositions via the Hadamard transform set up this interface. At least this serves as our apology for not making the super-classical power leap right off the page in chapter 11 and also explains the need to follow the linear algebra and the classical number-theoretic parts of the argument closely.

• *So where else can it be?* Perhaps Lucy is right—we fear that *everything* needs to be considered together to find the quantum power. The lever may have no single base. Rather, we may need an Archimedean lens by which to focus the ensemble of components we have shown. Perhaps the focus can be an *invariant* from algebraic geometry that is expressible via ideals of polynomials over some ring or field. An algebraic-geometric invariant is the source of what remain the only general super-linear lower bounds on arithmetical circuits for low-degree polynomial functions, which were obtained by Baur and Strassen (1982) following Strassen (1973). We suspect that this algebraic mechanism should have some natural role in metering quantum circuits as well.

If we step back from the deep difficulties of computational complexity theory and stay with subjects in communication and information such as quantum cryptography, then the power is easy to find, combining entanglement and interference and human–computer interaction. Although our emphasis on quantum algorithms has stood apart from these subjects' concerns, we hope the coverage here of quantum computing fundamentals has opened a way into them with increased amplitude. Whatever your path, provided it augments and does not cancel, we wish for it a measure of success.

Bibliography

Aaronson, S. (2005). Quantum computing, postselection, and probabilistic polynomial-time. *Proc. Royal Society A 461*, 3473–3482.

Aaronson, S. (2010). BQP and the polynomial hierarchy. In *Proceedings, 42nd Annual ACM Symposium on the Theory of Computation*, pp. 141–150.

Aaronson, S. (2013). *Quantum Computing Since Democritus*. Cambridge University Press.

Aaronson, S. and Arkhipov, A. (2010, November). The computational complexity of linear optics. Available at http://arxiv.org/abs/1011.3245.

Aaronson, S. and Gottesman, D. (2004). Improved simulation of stabilizer circuits. *Phys. Rev. A 70*, 052328.

Adleman, L., DeMarrais, J., and Huang, M.-D. (1997). Quantum computability. *SIAM J. Comput. 26(5)*, 1524–1540.

Aharonov, D. (1998). Quantum computation—a review. *Annual Review of Computational Physics VI, World Scientific*, 259–346.

Aharonov, D. (2003, January). A simple proof that Toffoli and Hadamard are quantum universal. Available at http://arxiv.org/pdf/quant-ph/0301040/.

Aharonov, D. (2008). Quantum computation. Available at http://arxiv.org/abs/quant-ph/9812037.

Aharonov, D., Landau, Z., and Makowsky, J. (2006). The quantum FFT can be classically simulated. Available at http://arxiv.org/abs/quant-ph/0611156.

Allender, E., Loui, M., and Regan, K. (2009a). Ch. 27: Complexity classes. In M. Atallah (Ed.), *CRC Handbook on Algorithms and Theory of Computation: 2nd ed.*, Volume 1: General Concepts and Techniques. Chapman and Hall/CRC Press.

Allender, E., Loui, M., and Regan, K. (2009b). Ch. 28: Reducibility and completeness. In M. Atallah (Ed.), *CRC Handbook on Algorithms and Theory of Computation: 2nd ed.*, Volume 1: General Concepts and Techniques. Chapman and Hall/CRC Press.

Allender, E., Loui, M., and Regan, K. (2009c). Ch. 29: Other complexity classes and measures. In M. Atallah (Ed.), *CRC Handbook on Algorithms and Theory of Computation: 2nd ed.*, Volume 1: General Concepts and Techniques. Chapman and Hall/CRC Press.

Ambainis, A. (2007). Quantum walk algorithm for element distinctness. *SIAM J. Comput. 37*, 210–239.

Ambainis, A., Childs, A., Reichardt, B., Špalek, R., and Zhang, S. (2010). Any and-or formula of size N can be evaluated in time $N^{1/2+o(1)}$ on a quantum computer. *SIAM J. Comput. 39*, 2513–2530.

Bach, E., Miller, G., and Shallit, J. (1984). Sums of divisors, perfect numbers, and factoring. In *Proceedings of the 16th Annual ACM Symposium on Theory of Computing*, pp. 183–190.

Ball, P. (2011). Physics of life: The dawn of quantum biology. *Nature 474*, 272–274.

Barenco, A., Bennett, C., Cleve, R., DiVincenzo, D., Margolus, N., Shor, P., Sleator, T., Smolin, J., and Weinfurter, H. (1995). Elementary gates for quantum computation. *Physical Review Letters 52(5)*, 3457–3467.

Barenco, A., Deutsch, D., Ekert, A., and Jozsa, R. (1995). Conditional quantum dynamics and logic gates. *Physical Review Letters 74(20)*, 4083–4086.

Baur, W. and Strassen, V. (1982). The complexity of partial derivatives. *Theoretical Computer Science 22*, 317–330.

Beals, R., Buhrman, H., Cleve, R., Mosca, M., and de Wolf, R. (2001). Quantum lower bounds by polynomials. *J. ACM 48(4)*, 778–797.

Belovs, A., Childs, A., Jeffery, S., Kothari, R., and Magniez, F. (2013). Time-efficient quantum walks for 3-distinctness. In *Proceedings of the 40th International Colloquium on Automata,*

Languages, and Programming (ICALP 2013), Volume 7965 of *Lect. Notes Comp. Sci.*, pp. 105–122. Springer-Verlag.

Benioff, P. (1982). Quantum mechanical models of Turing machines that dissipate no energy. *Physical Review Letters 48(23)*, 1581–1585.

Bennett, C. (1973). Logical reversibility of computation. *IBM Journal of Research and Development 6*, 525–532.

Bennett, C. and Wiesner, S. (1992). Communication via one- and two-particle oeprators on Einstein-Podolsky-Rosen states. *Phys. Rev. Lett. 69*, 2881.

Bennett, C. H., Bernstein, E., Brassard, G., and Vazirani, U. (1997). Strengths and weaknesses of quantum computing. *SIAM J. Comput. 26(5)*, 1510–1523.

Bennett, C. H., Brassard, G., CrŽpeau, C., Jozsa, R., Peres, A., and Wootters, W. (1993). Teleporting an unknown quantum state by dual classical and EPR channels. *Physical Review Letters 70*, 1895–1898.

Bernstein, E. and Vazirani, U. (1997). Quantum complexity theory. *SIAM J. Comput. 26(5)*, 1411–1473.

Boneh, D. and Lipton, R. (1996). Algorithms for black box fields and their application to cryptography. In *Proceedings of Crypto'96*, Volume 1109 of *Lect. Notes Comp. Sci.*, pp. 283–297. Springer-Verlag.

Bouwmeester, D., Pan, J.-W., Mattle, K., Eibl, M., Weinfurter, H., and Zeilinger, A. (1997). Experimental quantum teleportation. *Nature 390*, 575–579.

Brassard, G., Høyer, P., Mosca, M., and Tapp, A. (2000, May). Quantum amplitude amplification and estimation. Available at http://arxiv.org/abs/quantph/0005055.

Brassard, G., Høyer, P., and Tapp, A. (1998). Quantum counting. In *Proceedings of the 25th International Colloquium on Automata, Languages, and Programming*, Volume 1443 of *Lect. Notes. Comp. Sci.*, pp. 820–831. Springer-Verlag.

Buhrman, H., Cleve, R., and Wigderson, A. (1998). Quantum vs. classical communication and computation. In *Proceedings, 30th Annual ACM Symposium on the Theory of Computing*, pp. 63–68.

Cai, J.-Y., Chen, X., Lipton, R., and Lu, P. (2010). On tractable exponential sums. In *Proceedings of FAW 2010*, pp. 48–59.

Childs, A. (2013). Quantum walk search. Lecture notes for course CO781/CS867/QIC823. Available at http://www.math.uwaterloo.ca/ amchilds/teaching/w13/l13.pdf.

Childs, A., Cleve, R., Jordan, S., and Yonge-Mallo, D. (2009). Discrete-query quantum algorithm for NAND trees. *Theory of Computing 5*, 119–123. Available at http://arxiv.org/abs/quant-ph/0702160.

Childs, A. and Kothari, R. (2012). Quantum query complexity of minor-closed graph properties. *SIAM J. Comput. 41*, 1426–1450.

Childs, A. and van Dam, W. (2008). Quantum algorithms for algebraic problems. Available at http://arxiv.org/abs/0812.0380.

Coppersmith, D. (1994). An approximate Fourier transform useful in quantum factoring. Technical Report IBM Research Report RC19642, IBM T.J. Watson Research Center. Also http://arxiv.org/abs/quant-ph/0201067.

Dawson, C., Haselgrove, H., Hines, A., Mortimer, D., Nielsen, M., and Osborne, T. (2004). Quantum computing and polynomial equations over the finite field Z_2. *Quantum Information and Computation 5*, 102–112.

Dawson, C. and Nielsen, M. (2006). The solovay-kitaev algorithm. *Quantum Information and Computation 6*, 81–95. Available at http://arxiv.org/abs/quant-ph/0505030.

Deutsch, D. (1985). Quantum theory, the Church-Turing principle, and the universal quantum computer. In *Proceedings of the Royal Society of London*, Volume 400(1818) of *Series A*, pp. 97–117.

Deutsch, D. (1989). Quantum computational networks. In *Proceedings of the Royal Society of London*, Volume 425(1868) of *Series A*, pp. 73–90.

Deutsch, D. and Jozsa, R. (1992). Rapid solution of problems by quantum computation. In *Proceedings of the Royal Society of London*, Volume 438(1907) of *Series A*, pp. 553–558.

DiVincenzo, D. (1995). Two-bit gates are universal for quantum computation. *Physical Review A 51(2)*, 1015–1022.

Farhi, E., Goldstone, J., and Gutmann, S. (2008). A quantum walk algorithm for the Hamiltonian NAND tree. *Theory of Computing 4*, 169–190. Available at http://arxiv.org/abs/quant-ph/0702144.

Feynman, R. P. (1982). Simulating physics with computers. *International Journal of Theoretical Physics 21*, 467–488.

Feynman, R. P. (1985, February). Quantum mechanical computers. *Optics News 11*, 11–20.

Fortnow, L. and Rogers, J. (1999). Complexity limitations on quantum computation. *J. Comput. Sys. Sci. 59(2)*, 240–252.

Fredkin, E. and Toffoli, T. (1982). Conservative logic. *International Journal of Theoretical Physics 21*, 219–253.

Gerdt, V. and Severyanov, V. (2006). A software package to construct polynomial sets over Z_2 for determining the output of quantum computations. *Nuclear Instruments and Methods in Physics Research A 59*, 260–264.

Gottesman, D. (1998). The Heisenberg representation of quantum computers. Available at http://arxiv.org/abs/quant-ph/9807006.

Griffiths, R. and Niu, C.-S. (1996). Semiclassical Fourier transform for quantum computation. *Phys. Rev. Lett. 76*, 3228–3231.

Grover, L. (1996). Fast quantum mechanical algorithm for database search. In *Proceedings of the 28th Annual ACM Symposium on Theory of Computing (STOC '96)*, pp. 212–219.

Grover, L. (1997). Quantum mechanics helps in searching for a needle in a haystack. *Physical Review Letters 79(2)*, 325–328.

Hirvensalo, M. (2010). *Quantum Computing*. Springer-Verlag.

Holevo, A. S. (1973). Some estimates of the information transmitted by quantum communication channels. *Problems of Information Transmission 9*, 77–83.

Homer, S. and Selman, A. (2011). *Computability and Complexity Theory* (2nd ed.). Springer-Verlag.

Jeffery, S., Kothari, R., and Magniez, F. (2013). Nested quantum walks with quantum data structures. In *Proceedings of the 24th ACM-SIAM Symposium on Discrete Algorithms (SODA 2013)*, pp. 1474–1485.

Jozsa, R. (2006). On the simulation of quantum circuits. Available at http://arxiv.org/abs/quant-ph/0603163.

Kaye, P., Laflamme, R., and Mosca, M. (2007). *An Introduction to Quantum Computing*. Oxford University Press.

Kempe, J. (2003). Quantum random walks—an introductory overview. *Contemporary Physics 44*, 307–327. Available at http://arxiv.org/abs/quant-ph/0303081.

Kepner, J. and Gilbert, J. (2011). *Graph Algorithms in the Language of Linear Algebra*. SIAM.

Kitaev, A. Y., Shen, A., and Vyalyi, M. (2002). *Classical and Quantum Computation*, Volume 47 of *Graduate Studies in Mathematics*. Amer. Math. Soc. Press.

Kuttler, K. (2012). *Elementary Linear Algebra*. The Saylor Foundation.

Lecerf, Y. (1963). Machines de Turing réversibles. *Comptes Rendues de l'Academie Francaise des Sciences 257*, 2597–2600.

Lee, T., Magniez, F., and Santha, M. (2013). Improved quantum query algorithms for triangle finding and associativity testing. In *Proceedings of the 24th ACM-SIAM Symposium on Discrete Algorithms*, pp. 1486–1502.

Lenstra, H. (1983). Integer programming with a fixed number of variables. *Mathematics of Operations Research 8*, 538–548.

Lloyd, S. (1995). Almost any quantum logic gate is universal. *Physical Review Letters 75(2)*, 346–349.

Magniez, F., Nayak, A., Roland, J., and Santha, M. (2007). Search via quantum walk. In *Proceedings, 39th Annual ACM Symposium on Theory of Computing*, pp. 575–584.

Magniez, F., Nayak, A., Roland, J., and Santha, M. (2011). Search via quantum walk. *SIAM J. Comput. 40*, 142–164.

Marcikic, I., de Riedmatten, H., Tittel, W., Zbinden, H., and Gisin, N. (2003). Long-distance teleportation of qubits at telecommunications wavelengths. *Nature 421*, 509.

Markov, I. and Shi, Y. (2008). Simulating quantum computation by contracting tensor networks. *SIAM Journal on Computing 38*, 963–981.

Mosca, M. (2009). Quantum algorithms. In R. Meyers (Ed.), *Encyclopedia of Complexity and Systems Science*, pp. 7088–7188. Springer-Verlag. Available at http://arxiv.org/abs/0808.0369.

Nielsen, M. and Chuang, I. L. (2000). *Quantum Computation and Quantum Information*. Cambridge University Press.

Pippenger, N. and Fischer, M. (1979). Relationships among complexity measures. *Journal of the ACM 26*, 361–381.

Portugal, R. (2013). *Quantum Walks and Search Algorithms*. Springer-Verlag.

Preskill, J. (2004). Quantum information and computation: lecture notes for Physics 219. Available at http://www.theory.caltech.edu/~preskill/ph219/index.html#lecture.

Reichardt, B. (2011a). Faster quantum algorithm for evaluating game trees. In *Proceedings, 2011 ACM-SIAM Symposium on Discrete Algorithms*, pp. 546–559.

Reichardt, B. (2011b). Reflections for quantum query algorithms. In *Proceedings, 2011 ACM-SIAM Symposium on Discrete Algorithms*, pp. 560–569.

Reichardt, B. and Špalek, R. (2008). Span-program-based quantum algorithm for evaluating formulas. In *Proceedings, 40th Annual ACM Symposium on the Theory of Computing*, pp. 103–112.

Reingold, O. (2005). Undirected ST-connectivity in log-space. In *Proceedings, 37th Annual ACM Symposium on Theory of Computing*, pp. 376–385.

Rieffel, E. and Polak, W. (2011). *Quantum Computing: A Gentle Introduction*. MIT Press.

Salart, D., Baas, A., Branciard, C., Gisin, N., and Zbinden, H. (2008). Testing spooky action at a distance. *Nature 454*, 861–864.

Santha, M. (2008). Quantum walk based search algorithms. In *Proceedings, 5th Annual Conference on Theory and Applications of Models of Computation*, Volume 4978 of *Lect. Notes Comp. Sci.*, pp. 31–46. Springer-Verlag.

Savage, J. (1972). Computational work and time on finite machines. *Journal of the ACM 19*, 660–674.

Shi, Y. (2003). Both Toffoli and Controlled-NOT need little help to do universal quantum computation. *Quantum Information and Computation 3*, 84–92.

Shor, P. (1997). Polynomial-time algorithms for prime factorization and discrete logarithms on a quantum computer. *SIAM J. Comput. 26(5)*, 1484–1509.

Shor, P. W. (1994). Algorithms for quantum computation: Discrete logarithms and factoring. In *Proceedings of the 35th Annual IEEE Symposium on the Foundations of Computer Science*, pp. 124–134.

Simon, D. (1997). On the power of quantum computation. *SIAM J. Comput. 26(5)*, 1474–1483.

Sipser, M. (2012). *Introduction to the Theory of Computation* (3rd ed.). Cengage Learning.

Smolin, J., Smith, G., and Vargo, A. (2013). Oversimplifying quantum factoring. *Nature 499*, 163–165.

Strassen, V. (1973). Berechnung und Programm II. *Acta Informatica 2*, 64–79.

Szegedy, M. (2004). Quantum speed-up of Markov chain based algorithms. In *Proceedings of the 45th IEEE Symposium on Foundations of Computer Science*, pp. 32–41.

Toffoli, T. (1980). Reversible computing. In *Proceedings of the 7th Annual International Colloquium on Automata, Languages, and Programming*, Volume 85 of *Lect. Notes Comp. Sci.*, pp. 634–644. Springer-Verlag.

Turing, A. (1936). On computable numbers, with an application to the Entscheidungsproblem. *Proceedings of the London Mathematical Society 42*, 230–265.

Valiant, L. (1984). Short monotone formulae for the majority function. *Journal of Algorithms 5*, 363–366.

van den Nest, M. (2013). Efficient classical simulations of quantum Fourier transforms and normalizer circuits over Abelian groups. *Quantum Information and Computation 13*, 1007–1037.

Venegas-Andraca, S. (2012). Quantum walks: a comprehensive review. *Quantum Information Processing 11*, 1015–1116.

Williams, C. (2011). *Explorations in Quantum Computing*. Springer-Verlag.

Williams, V. and Williams, R. (2010). Subcubic equivalences between path, matrix, and triangle problems. In *Proceedings of the 51st Annual IEEE Symposium on Foundations of Computer Science*, pp. 645–654.

Yanofsky, N. and Mannucci, M. (2008). *Quantum Computing for Computer Scientists*. Cambridge University Press.

Yao, A.-C. (1993). Quantum circuit complexity. In *Proceedings of 34th Annual IEEE Symposium on Foundations of Computer Science*, pp. 352–361.

Index